W0042979

Journal of
Neural
Transmission

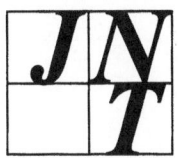

Supplementum 33

L. Deecke and P. Dal-Bianco (eds.)

Age-associated
Neurological Diseases

Springer-Verlag Wien New York

Prof. Dr. Lüder Deecke
Prof. Dr. Peter Dal-Bianco
Department of Neurology, University of Vienna, Austria

This work is subject to copyright.
All rights are reserved, whether the whole or part of the material is concerned, specifically those of translation, reprinting, re-use of illustrations, broadcasting, reproduction by photocopying machines or similar means, and storage in data banks.
© 1991 by Springer-Verlag Wien
Typeset by Best-set Typesetter Ltd, Hong Kong

Printed on acid-free paper

Product Liability: The publisher can give no guarantee for information about drug dosage and application thereof contained in this book. In every individual case the respective user must check its accuracy by consulting other pharmaceutical literature. The use of registered names, trademarks, etc. in this publication does not imply, even in the absence of a specific statement, that such names are exempt from the relevant protective laws and regulations and therefore free for general use.

With 30 Figures

ISSN 0303-6995
ISBN-13:978-3-211-82261-6   e-ISBN-13:978-3-7091-9135-4
DOI: 10.1007/978-3-7091-9135-4

Preface

The papers compiled in this supplementum are a selection of the best contributions presented at the 19th Central-European Neurological Symposium (CNS 19) held on June 29 – July 1, 1989 in Vienna. The main topic of this conference was degenerative and age-associated neurological diseases. In recent decades life expectancy has dramatically increased, at least in the industrialized countries. This has led to extreme distortions of the so-called population pyramids that no longer look like such but begin to almost resemble cylinders. As a consequence of this "overaging" of the population, diseases that are associated with age have become much more common than before. It was thus more than reasonable to devote a congress of the CNS series to these important neurological diseases.

The following fields of interest are covered: Age-associated memory impairment (AAMI), Alzheimer's and other dementias, Parkinson's disease and other movement disorders, stroke and others. Concerning the dementias, some papers deal with diagnosis employing neuro-imaging methods such as MRI, CT, PET and SPECT, others using electrophysiological methods. An important aspect in the early preclinical diagnosis of dementia is the inclusion of neuropsychological tests to enhance the chance of effective early treatment. Also drugs that are now under clinical investigation are discussed and preliminary results are presented. We hope that these efforts have succeeded in attracting all physicians, who are treating elder patients suffering from age-associated neurological diseases, to this volume and that they will benefit from it with its up to date contributions in one of the most important fields in curative and preventive neurology.

We would like to thank Prof. Dr. A. Carlsson, Managing Editor, Prof. Dr. P. Riederer, Coordinating Editor of "Journal of Neural Transmission", and Springer-Verlag Wien New York, especially Mr. Frank Chr. May.

L. DEECKE
P. DAL-BIANCO

Vienna, June 1991

Contents

Contents

Listed in Current Contents

J Neural Transm (1991) [Suppl] 33: 1–6
© by Springer-Verlag 1991

Diagnosis, assessment and treatment of age-associated memory impairment

T. H. Crook III and G. J. Larrabee

Memory Assessment Clinics Inc., Bethesda, Md, U.S.A.

Summary. We have reviewed diagnostic criteria and assessment procedures for AAMI, as well as pharmacologic and behavioral treatments for this condition. This research gives reason to hope that an important behavioral deficit associated with aging may be modified through drug and behavioral treatment.

Introduction

It is noteworthy that despite the extensive empirical evidence demonstrating decline in the ability to learn and remember certain types of information in the later decades of life (Fozard, 1985; Poon, 1985), there has been no generally accepted diagnostic classification for persons who experience such a decline. Some have applied the term "benign senescent forgetfulness" (BSF; Kral, 1962, 1966) to describe such persons, however specific diagnostic criteria have not been delineated. Moreover, BSF was originally introduced to describe borderline memory pathology in later life rather than "normal" age-related memory loss (Kral, 1962, 1966). Consequently, the diagnostic task in identifying older adults who have experienced memory decline associated with normal developmental processes is to distinguish these individuals from persons who 1) have experienced no such loss and 2) have memory loss characteristic of specific disease states such as Alzheimer's disease or Multi-infarct dementia.

The lack of an appropriate classification for "normal" age-related memory impairment has been recognized by the National Institute of Mental Health (NIMH) in the United States as an obstacle to clinical studies of memory loss in non-demented, older adults (Crook et al., 1986). The significance of such clinical studies is underscored by the large proportion of older adults who may experience memory loss and the dramatic growth of the elderly population. Of particular importance are investigations directed at managing, treating or preventing age-related memory deficits. Indeed, comparable deficits in both non-human primates and rodents are pharmacologically modifiable (Bartus and Dean, 1985) and also, may be modified both behaviorally and pharmacologically in humans (Crook, 1989; Crook and

Larrabee, 1988a; McEntee and Crook, 1990, 1991; Sheikh et al., 1986; West and Crook, 1990).

Diagnosis of age-associated memory impairment

In consideration of the significance of clinical studies of memory loss in non-demented older adults and the need for precise diagnostic criteria in such research, the NIMH formed a workgroup of experts to consider diagnostic criteria and terminology. The diagnostic term proposed by the NIMH workgroup was Age-Associated Memory Impairment (AAMI; Crook et al., 1986). The interested reader is referred to this original paper, as well as papers by Crook (1989) and Crook and Larrabee (1988a) for detailed specification of AAMI inclusion and exclusion criteria.

To summarize, the AAMI inclusion criteria apply to persons at least 50 years of age, with complaints of memory loss in everyday life (e.g., forgetting names of persons following introduction; misplacing objects), who perform at least one standard deviation below the mean level established for young adults on a standardized test of secondary (recent) memory such as the Benton Visual Retention Test (Benton, 1974) or the Logical Memory or Paired Associate Learning subtests of the Wechsler Memory Scale (WMS; Wechsler, 1945). Additionally, the subjects must have adequate intellectual function, determined by a scaled score of at least 9 on the Wechsler Adult Intelligence Scale (WAIS; Wechsler, 1955) Vocabulary subtest. Moreover, the subject should demonstrate absence of dementia, determined by a score of 24 or higher on the Mini-Mental State (Folstein et al., 1975; several investigators have chosen a score of 27 rather than 24 to exclude questionable cases of dementia: cf. Bleecker et al., 1988).

Exclusion criteria include delirium, confusion, or other disturbances of consciousness; any neurologic disorder that could produce cognitive deterioration (e.g., Alzheimer's disease, stroke); history of significant head trauma; history of inflammatory brain disease; current psychiatric diagnosis; alcohol or drug dependence or current use of psychotropic medication; and any systemic medical disorder which could produce cognitive deterioration (e.g., heart disease, hepatic disease). Exclusion criteria are determined by history, as well as by direct examination (e.g., Hamilton Depression Scale, Hamilton, 1960; modified Hachinski Ischemia Score, Rosen et al., 1980).

To date, these criteria have been applied in several clinical trials of candidate pharmaceutical compounds in the United States and Europe. A variety of neurotransmitter systems have been the focus of investigation (Crook and Larrabee, 1988a; Larrabee et al., 1990; McEntee and Crook, 1990, 1991).

Assessment of AAMI

In addition to issues of diagnostic classification, investigations concerning the occurrence, nature, and treatment of adult-onset memory loss have been hindered by limitations in assessment procedures. These limitations have included: 1) lack of adequate normative data, 2) failure to employ current cognitive paradigms, 3) lack of alternate forms, 4) absence of forms appropriate for different languages and cultures, and 5) lack of clinical relevance or "ecologic validity" (Crook et al., 1990; Cunningham, 1986; Larrabee and Crook, 1988a).

In consideration of these limitations, an effort was begun a decade ago to develop a series of technologically refined, clinically relevant, and "ecologically valid" performance tests for use in adult-onset cognitive disorders (Crook et al., 1979; Ferris et al., 1980). In 1985 a second generation of these measures was introduced utilizing state-of-the-art computer graphics and laser disk technology (Crook et al., 1986). These tests combine a high degree of face validity and simulation of everyday memory tasks (e.g, name-face learning, object location recall, facial memory, and grocery list learning) with current clinical and experimental paradigms such as delayed non-matching to sample, selective reminding, and signal detection (Larrabee and Crook, 1988a,b). Previous factor analytic research on this battery has demonstrated multi-dimensional structure, including factors of general (verbal and visual) everyday memory, attention, vigilance, psychomotor speed, everyday verbal and everyday visual memory (Crook and Larrabee, 1988b; Larrabee and Crook, 1989a). Cluster analyses have also demonstrated several different subtypes of persons with differential everyday memory strengths and weaknesses (Larrabee and Crook, 1989b).

Performance on one of the computerized everyday memory tasks, Name-Face Association, is displayed in Fig. 1. This test measures a person's ability to learn the first names, following introduction, of 14 persons (actors), presented by live video recording stored on laser disk, over three acquisition trials (presentation order is randomized each trial), with a 40-minute delayed recall. The data in Fig. 1, based on 1,976 healthy subjects, aged 18 to over 70, graphically illustrate the striking decline in name-face memory performance

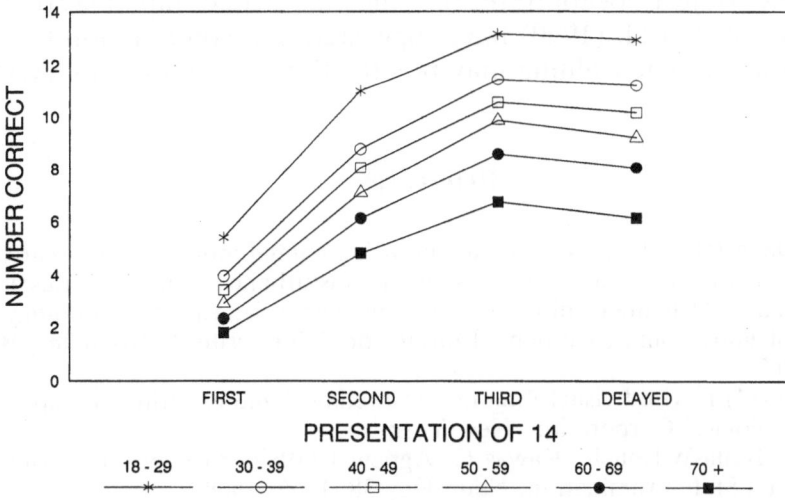

Fig. 1. Performance differences among age groups on the name-face association test

that occurs across the adult life span. This decline begins in the 30s and proceeds in a generally continuous linear fashion through the next 4 decades of life.

These data are particularly informative when considered in terms of the psychometric criteria for AAMI. The normative data for Name-Face Memory (Crook et al., 1990) allow calculation of the percentage of persons of different age decades meeting the AAMI criterion for secondary (recent) memory performance. These percentages are quite sizeable; for example, 70% of 50–59 year-old, 79% of 60–69 year old, and 91% of over 70 year-old subjects score at least one standard deviation below the mean performance of 18 to 29 year-old subjects on Name-Face 40-minute delayed recall.

Treatment of AAMI

Since the proposal of the AAMI criteria (Crook et al., 1986), several clinical drug trials have been initiated, directed at modifying the memory disorder of AAMI. Pharmacologic treatment strategies covering a wide range of neuro-transmitter systems have been reviewed in detail in previous papers (Crook and Larrabee, 1988a; McEntee and Crook, 1990, 1991; Larrabee et al., 1990). Recently, Crook and colleagues (Crook et al., in press) reported a positive treatment response in a double-blind, placebo-controlled investigation of the effects of phosphatidylserine (BC-PS) in persons diagnosed with AAMI. Analysis of clinical subgroups suggested that persons within the sample who performed at a relatively low level prior to treatment were most likely to respond to BC-PS. These data suggest that BC-PS and other related phospholipids may prove beneficial in the treatment of AAMI.

Behavioral treatments have also been used with some success in persons with AAMI. Sheikh et al. (1986) reported significant improvement in name-face memory following a visual interactive imagery mnemonic. West and Crook (1990) found significant gains in name-face memory, grocery list learning, and object location recall following video tape-based imagery training. Larrabee et al. (1990) have suggested that combination treatments of drug and mnemonic training may be effective in ameliorating AAMI.

References

Bartus RT, Dean RL (1985) Developing and using animal models in the search for an effective treatment for age-related memory disturbances. In: Gottfries CG (ed) Normal aging, Alzheimer's disease, and senile dementia: aspects on etiology, pathogenesis, diagnosis, and treatment. Editions de l'Universite de Bruxelles, Brussels, pp 231–267

Benton AL (1974) Revised visual retention test: clinical and experimental applications. The Psychological Corporation, New York

Bleecker ML, Bolla-Wilson K, Kawas C, Agnew J (1988) Age-specific norms for the Mini-Mental State examination. Neurology 38: 1565–1568

Crook TH (1989) Diagnosis and treatment of normal and pathologic memory impairment in later life. Semin Neurol 9(1): 20–30

Crook TH, Bartus RT, Ferris SH, Whitehouse P, Cohen GD, Gershon S (1986) Age-associated memory impairment: proposed diagnostic criteria and measures of clinical change — Report of a National Institute of Mental Health workgroup. Dev Neuropsychol 2(4): 261–276

Crook T, Ferris S, McCarthy M (1979) The misplaced objects task: a brief test for memory dysfunction in the aged. J Am Geriatr Soc 27: 284–287

Crook TH, Johnson BA, Larrabee GJ (1990) Evaluation of drugs in Alzheimer's disease and age-associated memory impairment. In: Benkert O, Maier W, Rickels K (eds) Methodology of the evaluation of psychotropic drugs. Springer, Berlin Heidelberg New York Tokyo, pp 37–55

Crook TH, Larrabee GJ (1988a) Age-associated memory impairment: diagnostic criteria and treatment strategies. Psychopharmacol Bull 24(4): 509–514

Crook TH, Larrabee GJ (1988b) Interrelationships among everyday memory tests: stability of factor structure with age. Neuropsychology 2: 1–12

Crook TH, Larrabee GJ, Youngjohn JR (1990) Diagnosis and assessment of age-associated memory impairment. Clin Neuropharmacol 13 [Suppl 3]: 81–91

Crook TH, Salama M, Gobert J (1986) A computerized test battery for detecting and assessing memory disorders. In: Bes A, Cahn J, Hoyer S, Marc-Vergnes JP, Wisniewski HM (eds) Senile dementias: early detection. Libbey, London Paris, pp 79–85

Crook TH, Tinklenberg J, Yesavage J, Petrie W, Nunzi MG, Massari DC (1991) Effects of phosphatidylserine in age-associated memory impairment. Neurology (in press)

Cunningham WR (1986) Psychometric perspectives: validity and reliability. In: Poon LW, Crook T, Davis KL, Eisdorfer C, Gurland BJ, Kaszniak AW, Thompson LW (eds) Handbook for clinical memory assessment of older adults. American Psychological Association, Washington DC, pp 27–31

Ferris SH, Crook T, Clark E, McCarthy M, Rae D (1980) Facial recognition memory deficits in normal aging and senile dementia. J Gerontol 35: 707–714

Folstein MF, Folstein SE, McHugh PR (1975) Mini-Mental State: a practical method for grading the cognitive state of patients for the clinician. J Psychiatr Res 12: 189–198

Fozard JL (1985) Psychology of aging-normal and pathological age differences in memory. In: Brockelhurst JC (ed) Textbook of geriatric medicine and gerontology. Churchill Livingstone, Edinburgh, pp 122–144

Hamilton M (1960) A rating scale for depression. J Neurol Neurosurg Psychiatry 23: 56–62

Kral VA (1962) Senescent forgetfulness: benign and malignant. Can Med Assoc 86: 257–260

Kral VA (1966) Memory loss in the aged. Dis Nerv Syst 27: 51–54

Larrabee GJ, Crook TH (1988a) Assessment of drug effects in age-related memory disorders:clinical, theoretical, and psychometric considerations. Psychopharmacol Bull 24(4): 515–522

Larrabee GJ, Crook TH (1988b) A computerized everyday memory battery for assessing treatment effects. Psychopharmacol Bull 24(4): 695–697

Larrabee GJ, Crook TH (1989a) Dimensions of everyday memory in age-associated memory impairment. Psychological Assessment: A Journal of Consulting and Clinical Psychology 1(2): 92–97

Larrabee GJ, Crook TH (1989b) Interrelationships among everyday memory tests: stability of factor structure with age. Neuropsychology 2: 1–12

Larrabee GJ, McEntee WJ, Crook TH (1991) Age-associated memory impairment. In: Gamzu ER, Moos WH, Thal LJ (eds) Cognitive disorders: pathophysiology and treatment. Dekker, New York (in press)

McEntee WJ, Crook TH (1990) Age-associated memory impairment: a role for catecholamines. Neurology 40: 526–530

McEntee WJ, Crook TH (1991) Serotonin, memory, and the aging brain. Psychopharmacoloy 103: 143–149

Poon LW (1985) Differences in human memory with aging: nature, causes, and clinical implications. In: Birren JE, Schaie KW (eds) Handbook of the psychology of aging, 2nd edn. New York, pp 427–462

Rosen WG, Terry RD, Fuld PA, Katzman R, Peck A (1980) Pathological verification of ischemic score in differentiation of dementias. Ann Neurol 7: 486–488

Sheikh JI, Hill RD, Yesavage JA (1986) Long-term efficacy of cognitive training for age-associated memory impairment: a six-month follow-up study. Dev Neuropsychol 2: 413–424

Wechsler DA (1945) A standardized memory scale for clinical use. J Psychol 19: 87–95

Wechsler DA (1955) Manual for the Wechsler Adult Intelligence Scale. Psychological Corporation, New York

West RL, Crook TH (1991) Video training of imagery for mature adults. J Appl Cogn Psychol (in press)

Authors' address: T. H. Crook III, Ph.D., Memory Assessment Clinics, Inc., 8311 Wisconsin Avenue, Bethesda, Maryland 20814, U.S.A.

J Neural Transm (1991) [Suppl] 33: 7–12
© by Springer-Verlag 1991

Physiology of short-term verbal memory

**A. Starr[1], H. Pratt[1], H. Michalewski[1], J. Patterson[1], G. Barrett[2], F. Swire[2]*,
L. Deecke[3], D. Cheyne[3], R. Kristova[3], and G. Lininger[3]**

[1]University of California, Irvine, Ca, U.S.A., [2]National Hospital for Nervous Disease,
London, United Kingdom, and [3]University of Vienna, Austria

Summary. These studies document a series of brain events accompanying short-term memory functions. For auditory verbal material the sequence involves at least two different sites within auditory cortex subserving sensory and cognitive processes of memorization. During the scanning of the short-term store structures within the medial temporal lobes, presumably the hippocampus, are active. There is an inconsistency between these results and the clinical observations of the need for an intact dominant parietal lobe for auditory short-term memory to function normally. Magnetic recordings showed no focal dipolar source of activity in the parietal lobe during any aspect of auditory short-term memory. The discrepancy could be accounted for by considering the parietal lobe lesion as "disconnecting" the lateral temporal cortex from the deep medial hippocampal structures thereby impeding auditory short-term functions (Geschwind, 1965).

These studies show that the physiological analysis of brain events in the msec range can provide information about relatively complex cognitive processes underlying short-term memory. The magnetic and electrical recording methods provide a noninvasive way to study human brain functions involved in cognition that can then be correlated with behavioral measures of specific cognitive activities.

Short-term verbal memory is essential for our daily activities. For instance, remembering a new phone number, new names, or a list of groceries all engage short-term memory processes that we all wished functioned better. There is a wealth of experimental data about short-term verbal memory that has clarified the workings of this type of cognitive activity. The short-term store has a limited capacity of approximately 7 items with the auditory store being larger than the visual store. Exposure to interfering stimuli can impair short-term verbal memory (Peterson and Peterson, 1959). Reaction time measures show that the length of time needed to scan the short-term store increases linearly as the number of items memorized increases (Sternberg, 1969). Brain lesions in man localized to specific sites can disassociate short-

---

* Perished in the Pan Am tragedy at Lockerby, 1988

term and long-term memory processes. Lesions of the left angular gyrus of the parietal lobe are associated with disordered auditory short-term verbal memory (Warrington and Shallice, 1969) but are without effect on long-term memories. In contrast, lesions of the hippocampus bilaterally are associated with disordered long-term but not short-term verbal memory functions (Starr and Philips, 1970). Over the past five years I have worked with colleagues at Irvine, London, and Vienna using electrophysiological and magnetic recording methods in normals and in patients with brain lesions to help elucidate brain activity during short-term memory. The results show a sequence of events in different brain regions during memorizing and scanning of the short-term store. These events obey rules that differ from those defined by reaction time measures. This paper summarizes the results of our studies.

Methods

Brain potentials were recorded from at least three midline electrodes (Fz, Cz, and Pz) referenced to linked ears. Eye movements were also monitored so that trials in which they occurred could be excluded from the averaging process. In separate experiments, magnetic fields were measured from up to 63 sites over the skull. The details of the recording parameters and methods of analysis can be found in two of our publications (Pratt et al., 1989; Starr et al., 1990). The short-term memory processes were engaged in a systematic manner using a variant of a paradigm described by Sternberg (1969). A list of items to be memorized were presented once every 1.2 sec. followed in 2 sec. by a probe item (Fig. 1A) that had an equal probability of being (an "inset" probe) or not being (an "out-of-set" probe) a member of the immediately preceding memory set. The subject indicated his/her classification of the probe by responding as soon as possible by pressing one of two buttons and accuracy and reaction times were recorded. The items were both verbal (numbers 1 to 9) and nonverbal (musical notes from middle C to D, one octave higher) presented in the auditory or visual modality. Blocks of twenty trials were presented containing the same number of items (one, three, or five), the same type of item (verbal or nonverbal), and in the same modality (visual or auditory).

Results

Normal subjects

Reaction time measures increased in a linear fashion as the number of items memorized increased. The slopes were approximately 50 msec/item for verbal items (Fig. 1D) and twice that, 100 msec/item, for musical notes. There were no significant differences in reaction times between inset and out-of-set probes.

The brain potentials accompanying the probes consisted of early sensory events (N100, P200) that differed in scalp distribution and amplitudes between auditory and visual modes of presentation. A late sustained positivity peaking at a modal latency of 450 msec of largest amplitude parietally accompanied the presentation of all of the different probes (Fig. 1A). We believe this potential reflects brain activities related to the scanning of the

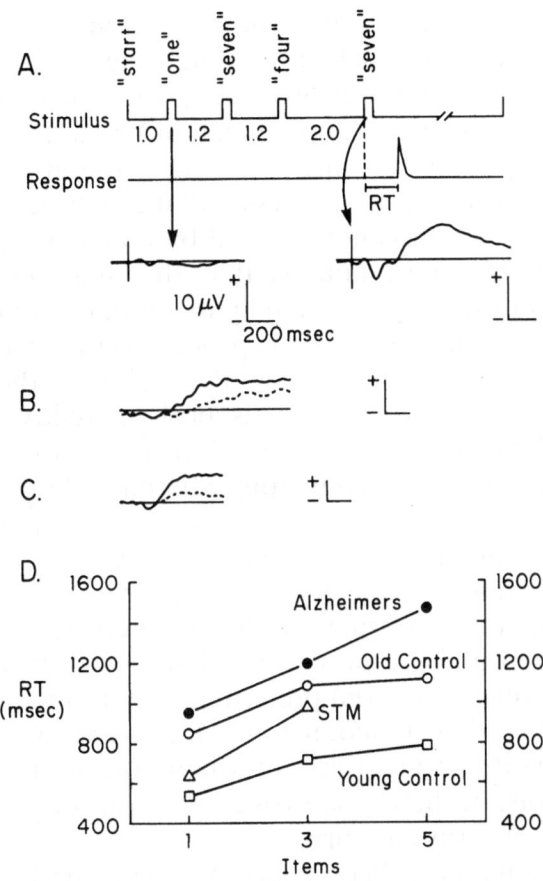

Fig. 1. A A schematic representation of the short-term memory task used to measure brain potentials and behavior. For the memory items there is a small negative potential shift over the frontal region (left trace). For the probe there is a large positivity seen over the parietal region (right trace). **B** Brain potentials to probes in normals with (dashed line) and without (solid line) an intervening interfering auditory task. Note the attenuating effect of interference. **C** Brain potentials to probes in subjects normal (solid line) and in one patient with disordered auditory verbal short-term memory from a lesion of the left angular gyrus. Note the attenuation of the potential in the patient. **D** RT as a function of set size for young normals, old normals, patients with disordered auditory short-term memory from a lesion of the dominant angular gyrus (STM), and patients with Alzheimer's disease

contents of the short-term store. The latency of this component increased with set size at approximately a 30 msec/item rate, approximately half that defined for reaction times. Moreover the increase in latency of the parietal positivity with increases in set size was similar for all of the various types of memory items, differing again from reaction times.

There were features of this parietal positivity that could be correlated with short-term memory processes. First, the parietal positivity differed for inset and out-of-set probes: the former began at an earlier latency than to the out-of-set probes. This finding is compatible with the idea that scanning of the contents of the short-term store can be completed sooner for inset than for out-of-set probes. For instance, if scanning of the short-term store followed a serial or even a random order then a match would be made for inset probes, on the average, halfway through the list. In contrast, a probe could be correctly classified as out-of-set only after the entire contents of the short-term store were scanned. Since reaction times do not differ between inset and out-of-set probes it is likely that the parietal positivity reflects an early stage in short-term memory processes that can distinguish between inset and out-of-set probes which is not realized by reaction time measures.

A second observation of note is that the amplitude of the parietal positivity to the inset probes is affected by its serial position within the list of items. Amplitudes were larger when the probe was the first or last item of the memorized list compared to the items in the middle of the list (Patterson et al., 1991). This serial position effect is both modality and item specific occurring only with verbal items presented in the auditory mode.

Finally, interference from competing stimuli, which is well known to decrease retrieval from the short-term store, is accompanied by a marked attenuation of the parietal positivity and a lengthening of the reaction time. These latter observations were made by Flora Swire when she had subjects perform an arithmetic task during the interval between the presentation of the last memory item and the probe. The late positivity was attenuated by more than half for verbal material presented in both the auditory and visual modalities (Fig. 1B) while reaction times increased by approximately 100 msec. These observations are relevant when considering the mechanisms underlying alterations of the probe evoked potentials that occurs in patients with disordered short-term memory.

Up to this point we have been concerned with the brain events accompanying the scanning of the short-term store. During memorization of the list of items a sustained low amplitude component appeared that was positive for visually presented digits and acoustically presented notes but negative or absent to acoustically presented digits. Recording from the frontal region showed that all items regardless of modality or type showed a progressive tendency to become negative as memorization progressed from the first to the fifth item in a list. Thus, during memorization there is a low amplitude potential, particularly to acoustically presented material, that is largest in the frontal derivations (Fig. 1A).

Patients

Two groups of patients have been studied: three patients with lesions of the left angular gyrus and impaired auditory short-term memory (Starr and Barrett, 1987) and twenty-five patients with a variety of dementing disorders. The first group of patients with disordered auditory short-term memory showed a selective attenuation or even absence of the parietal positivity to verbal probes presented in the auditory but not in the visual modality. The scanning rates of only the auditory short-term store, using their reaction times, was abnormal at 150 msec/item, almost three-fold greater than found in normals (Fig. 1D). The attenuation of the amplitude of the parietal positivity (Fig. 1C) was similar to that, found by Flora Swire, in normal subjects engaged in an interfering task. It has been suggested that interference affects auditory verbal short-term memory functions by preventing "rehearsal" using articulatory processes. One does not have to repeat the items "out loud" for rehearsal to occur since rehearsal can occur through activation of neural representations of the phonetic output system. It may be that patients with the selective disorder of auditory verbal short-term memory have alterations in their articulatory rehearsal capabilities leading to deficits similar to that seen in normals exposed to interfering stimuli. The patients do not have a disorder of their short-term memory functions, per se, since verbal material presented in the visual modality can be successfully stored and retrieved and the accompanying parietal positivities were of normal amplitude.

In patients with dementing illness reaction time measures of the time required to scan the contents of the short-term store increased two fold compared to aged matched controls (Fig. 1D). In contrast, the amplitude and latency of the parietal positivity were unaffected. Thus, in dementia, short-term memory processes represented by the amplitude and latency of the parietal positivity accompanying the classification of the probes appears normal. It is the use of this information for behavior (i.e., accuracy and reaction times) that becomes abnormal.

In both groups of patients the long latency potentials accompanying the memorization of the list of items have not yet been analyzed. However, the short latency sensory events (N100) were no different than normal.

Localization of brain structures that are active during short-term memory processes

The distribution of magnetic fields over the scalp during scanning of the auditory short store at the time of the parietal positivity to the probe (350–700 msec), showed outward fluxes over the left hemisphere and inward fluxes over the right hemisphere. These findings are compatible with activity arising from two dipolar sources localized deep in brain in the region of the medial temporal lobes. During memorization, at the time of the sustained

Table 1. Localization of brain activity during auditory processing of a verbal item

Memorization
100 msec — Sensory processing in primary auditory cortex, Heschl's gyrus
200–600 msec — Cognitive processes of attention or short-term memory, temporal cortex adjacent to Heschl's gyrus

Retrieval
100 msec — Sensory processing in primary auditory cortex, Heschl's gyrus
200–800 msec — Cognitive processes of memory scanning, medial temporal lobes
— Hippocampus

frontal negativity to acoustically presented digits (300–500 msec), activity could be localized to a dipole source in the region of auditory cortex adjacent to Heschl's gyrus. Even earlier, at 100 msec after the items' appearance, activity originating from a dipolar source in Heschl's gyrus was clearly evident. Thus a shifting sequence of activity could be defined accompanying the engagement of auditory short-term memory (Table 1): at 100 msec there is activity in primary auditory sensory cortex; at 300–500 msec, during the presentation of the items to be memorized, there is activity in auditory cortex adjacent to primary sensory cortex; finally at 300–700 msec during the presentation of probes and the scanning of the short-term store, there is activity in the medial portions of both temporal lobes.

Acknowledgement

Supported in part by NIH Grant #11876.

References

Geschwind N (1965) Disconnection syndromes in animal and man, part II. Brain 88: 585–644

Patterson JV, Pratt H, Starr A (1991) Event-related potential correlates of the serial position effect in short-term memory. Electroencephalogr Clin Neurophysiol (in press)

Peterson LR, Peterson M (1959) Short-term retention of individual verbal items. J Exp Psychol 58: 193–198

Pratt H, Michalewski HJ, Barrett G, Starr A (1989) Brain potentials in a memory-scanning task. I. Modality and task effects in potentials to the probes. Electroencephalogr Clin Neurophysiol 72: 407–421

Sternberg S (1969) Memory scanning: mental processes revealed by reaction-time experiments. Am Sci 4: 421–457

Starr A, Barrett G (1987) Disordered auditory short-term memory in man and event-related potentials. Brain 110: 935–959

Starr A, Cheyne D, Kristeve R, Linninger G, Deecke L (1991) Localization of brain activity during auditory short-term memory. Brain Res (submitted)

Starr A, Phillips L (1970) Verbal and motor memory in the amnestic syndrome. Neuropsychologia 8: 75–88

Warrington EK, Shallice T (1969) The selective impairment of auditory short-term memory. Brain 92: 885–896

Authors' address: Dr. A. Starr, University of California Irvine, Irvine, California 92717, U.S.A.

J Neural Transm (1991) [Suppl] 33: 13–19
© by Springer-Verlag 1991

Positron emission tomography in the differential diagnosis of organic dementias

W.-D. Heiss, J. Kessler, B. Szelies, M. Grond, G. Fink, and **K. Herholz**

Max-Planck-Institut für neurologische Forschung and Klinik für Neurologie, Köln, Federal Republic of Germany

Summary. At present, PET is the only technology affording the quantitative, three-dimensional imaging of various aspects of brain function. Since function and metabolism are coupled, and since glucose is the dominant substrate of the brain's energy metabolism, studies of glucose metabolism by PET of 2(18F)-fluoro-2-deoxy-D-glucose (FDG) are widely applied for investigating the participation of various brain systems in simple or complex stimulations and tasks. In focal or diffuse disorders of the brain, functional impairment of affected or inactivated brain regions is a reproducible finding.

While glucose metabolism is decreased slightly with age in a regionally different degree, in most types of dementia severe changes of glucose metabolism are observed. Degenerative dementia of the Alzheimer type is characterized by a metabolic disturbance most prominent in the parieto-occipito-temporal association cortex and later in the frontal lobe, while primary cortical areas, basal ganglia, thalamus, and cerebellum are not affected. By this typical pattern Alzheimer disease can be differentiated from other dementia syndromes, as e.g., Pick's disease (with the metabolic depression most prominent in the frontal and temporal lobe), multi infarct dementia (with multiple focal metabolic defects), and Huntington's chorea (with metabolic disturbance in the neostriatum). In demented patients PET studies can also be applied to the quantification of treatment effects on disturbed metabolism.

The clinical diagnosis of Alzheimer's disease (AD) is usually based on progressive dementia occurring in middle or late life, and by exclusion of other diseases, that could have caused a cognitive decline or a personality change (McKhann et al., 1984). Dementias, which are clinically manifested primarily as non-localizable disturbances of cerebral function, can hardly be diagnosed by conventional supplementary neurological investigations, which detect mainly localized morphologic lesions. The primary (degenerative) diseases leading to dementia are accompanied by atrophic changes in the brain visible at CT only in the late stages. Progressive cell loss and reduced cell and synaptic activity lead to a reduction of metabolism and blood flow which can be visualized with the aid of functional imaging techniques. Since glucose is the most important substrate of cerebral energy metabolism, studies of glucose metabolism are currently the best method of detecting and quantifying functional disturbances of the brain. The glucose metabolic rate can be determined regionally and three-dimensionally in the brain by means of positron emission tomography.

Glucose metabolism in healthy subjects

The different rates of glucose metabolism in various regions of the brain depending on their functional activity have been determined in a number of studies (Heiss et al., 1984). The mean glucose turnover rate of 42 normal subjects (age 43 ± 19.1 years, 14 women, 28 men) was 34.6 ± 3.83 μmol/100 g cerebral tissue/min. Highly significant regional differences (Fig. 1) with values between 40 and 50 μmol/100 g/min were detected in the

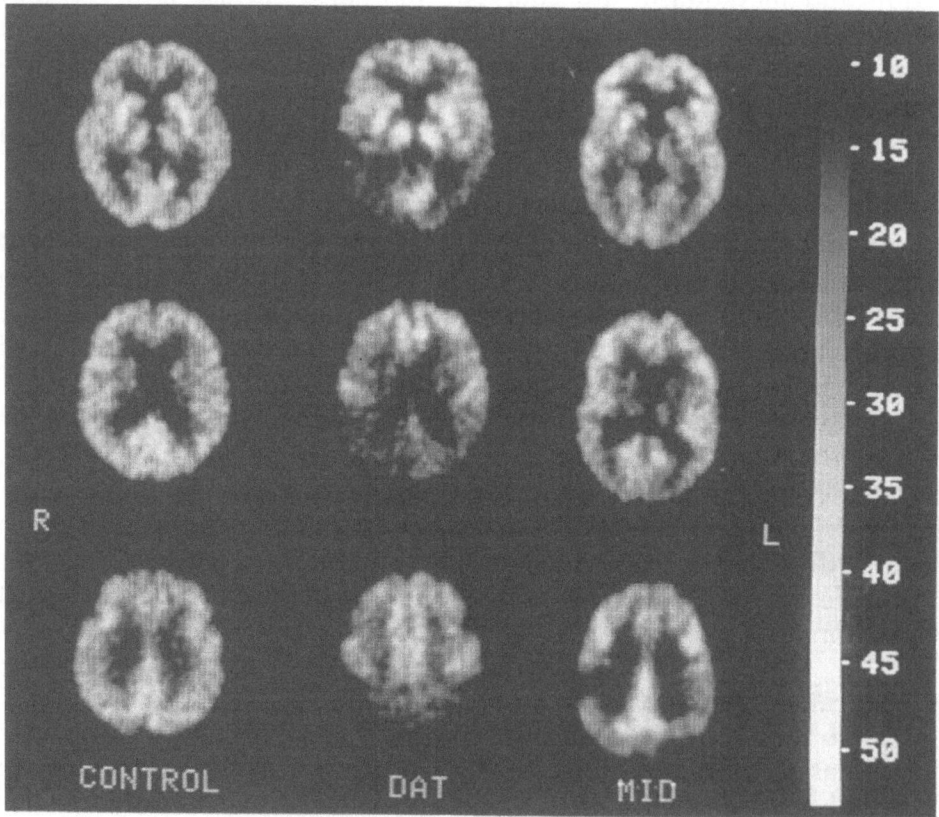

Fig. 1. PET images of glucose metabolism as determined by FDG (values in μmol/100g/min according to gray scale) in three levels (thalamus/basal ganglia, corpus nuclei caudati centrum semiovale) of a normal volunteer (left), in a patient with Alzheimer dementia (middle) and a case with multi infarct syndrome (right). In the AD patient CMRGlu is typically reduced in the parieto-temporo-occipital and the frontal association cortex with normal values in primary cortical fields and the basal ganglia. The MID case shows focally reduced CMRGlu within the ischemic lesions (right parietal, left occipital)

striatum, upper limbic system, insula, frontal cortex and primary visual cortex, between 35 and 40 µmol/100 g/min in the other gray structures of the hemispheres, between 30 and 35 µmol/100 g/min in the cerebellum and hippocampal structures and below 20 µmol/100 g/min in the medullary layer. Our studies confirm additionally a certain age dependency: the global rate of cerebral glucose metabolism showed a decline with advancing age which, although statistically significant ($P < 0.05$) represented less than 2% per decade.

Metabolic disturbances in dementia syndromes

Dementias are a very heterogeneous group of diseases, common to all of which are deterioration of intellectual function and memory and a degeneration of the personality which are clearly differentiable from the usual age-related changes — slight forgetfulness, change in the intelligence structures — and greatly exceed them in scope. Clinical symptoms include impairment of learning ability and memory and reduction of attention, orientation, critical faculty and judgement. These are frequently associated with disturbances of visual-spatial orientation, speech functions and apraxie.

Primary degenerative dementias

Primary degenerative dementias of the Alzheimer type (AD) account for more than 50% of all dementia disorders. Patients with AD show a reduction of cerebral glucose metabolism proportional to the severity of the dementia, in which the reduced metabolism is detectable before the occurrence of atrophic changes in CT and shows significant regional differences: the bilateral local reductions are especially pronounced in the parieto-temporal and frontal cortex (Fig. 1) and do not affect the primary visual and sensorimotor cortex or the subcortical structures and the cerebellum (Kuhl et al., 1983; DeLeon et al., 1983; Friedland et al., 1983; Duara et al., 1986).

Of all the regional values of glucose metabolism those in the parieto-temporal cortex correspond best to the severity of mental impairment while rCMRGl in primary cortical areas (e.g., sensorimotor cortex,) is not related to severity or duration of AD.

In the second, but much rarer form of primary degenerative dementia, Pick's disease, the first and most marked metabolic changes — in analogy to the primary localization of pathological changes — are seen in the frontal and temporal lobe (Szelies and Karenberg, 1986). This distinctly different pattern of damage allows Pick's disease to be differentiated from AD.

The pattern of metabolic disturbance is also characteristic in Huntington's chorea which in addition to the extrapyramidal-hyperkinetic syndrome is always accompanied by dementia disorders. The glucose turnover rate in the neostriatum is already significantly reduced in the early stages of this disease and as the severity and duration of the disease increases, metabolism is seen to be reduced in the nucleus caudatus and putamen, and later (according to the degree of severity of dementia) also the cerebral cortex (Kuhl et al., 1984).

In Parkinson's disease, a degeneration of the dopaminergic nigrostriatal system, glucose metabolism is usually not altered, in contrast to the reduction of the dopaminergic endings in the basal ganglia demonstrated by means of PET of ^{18}F-dopa (Nahmias et al., 1985). Only on development of a dementia, a frequent concomitant of Parkinson's disease, are the metabolic changes typical of AD also in evidence (Kuhl et al., 1985).

Vascular dementias

Focal cerebral lesions caused by blood flow disturbances can induce dementia syndromes through two mechanisms in particular: Multi-infarct dementias (MID) together with the AD-MID mixed forms account for about 30% of all dementia syndromes. A clinical differentiation on the basis of rating scales (Hachinski et al., 1975) is often difficult, and diagnostic classification is often easier on the basis of morphological lesions demonstrated by CT or MRI. In MID patients, PET can clearly differentiate mostly multilocular metabolic reductions (Fig. 1) from the pattern typical of AD (Kuhl et al., 1983). Detection of ischemic lesions in the medullary layer in MID and Binswanger's disease can be performed with great sensitivity by means of T_2-weighted MRI and the regions of reduced metabolism then correspond to the superjacent deafferentated cortical areas.

Small isolated infarctions in the cerebral regions which are particularly important for the integrity of the personality lead to disturbances of behaviour, affect, mood and intellectual performance. This is especially true of infarcts in the area supplied by the anterior cerebral artery, but also for small localized infarcts in strategically important regions, for example unilaterally in the anterior centre of the thalamus or bilaterally in the median thalamus: they also lead to permanent cognitive and amnestic losses.

Dementias of other etiology

Various other causes — inflammatory diseases, as herpes encephalitis, HIV encephalopathy, Creutzfeldt-Jakob disease, posttraumatic encephalopathy, toxic affections of the central nervous system, communicating hydrocephalus — can lead to severe dementias which then must be differentiated from the more frequent etiologies. Their representation in PET is usually not following a typical pattern — as AD or Pick's disease — or metabolic changes are not restricted to small areas — as in postinfarction states. Therefore, a differentiation from the more frequently occurring form of dementia — e.g., AD and MID — can usually be achieved.

It is often difficult to differentiate the affective disorders with impairment of drive and psychomotoricity seen in depressions from similar symptoms in the early stages of dementia. The metabolic patterns observed in depressed patients are not comparable to the characteristic changes seen in dementias, particularly in AD. When the overall metabolic level is in relation to the mood (Baxter et al., 1987) there are sometimes regional differences of varying distribution; a pattern typical of depression or correlations between the metabolic values of certain regions with the severity of specific symptoms or function deficits have not so far been described.

Differentiation of AD from other dementias

In order to test the value of FDG-PET in the differential diagnosis of dementias, rCMRGl measurements of 19 patients with probable Alzheimer's disease according to the NINCDS-ARDA criteria (McKhann et al., 1984) were compared to those from 19 age-matched healthy subjects and 22 patients with cognitive impairment due to other diseases. In comparison to the 19 healthy normals the AD patients (age 49 to 71 years, cognitive deterioration for 1–5 years, average mini mental status 14.5 ± 7.3, average global deterioration scale 4.9 ± 0.9) were characterized by significantly lower global cerebral metabolic rates for glucose (30.3 ± 3.2 µmol/100 g/min) and considerable metabolic asymmetries with the most conspicious decrease of regional CMRGl (in terms of Z-transforms) found in the supramarginal and angular gyrus, the adjacent parts of the superior temporal gyrus and the medial frontal gyrus. While abnormally low metabolism ($Z < -2$) in supramarginal/angular gyrus was observed in all AD patients and is a highly sensitive indicator of the disease, abnormal metabolism in temporal and frontal association areas supports the diagnosis, but is not mandatory.

However, despite the decrease of global CMRGl in AD, some regions maintained a strikingly normal metabolism even in most severely affected patients. Those regions were the cerebellum, brainstem, primary sensori-motor cortex and the occipital cortex including the cortex around the calcarine fissure.

Considering the contrast between typically affected and non-affected regions, a ratio R of the rCMRGl in those regions was calculated. Its average value in normals was 1.09 (SD 0.084, range 1.01 to 1.15) vs. 0.77 (SD 0.11, range 0.60 to 1.00), thus a complete separation of the two groups was achieved (Fig. 2).

The diagnostic criteria of AD deducted from the metabolic pattern on FDG-PET — reduced temporo-parietal metabolism, normal cerebellar metabolism, reduced CMRGl ratio of temporo-parietal and frontal association areas to primary sensory areas, brainstem and cerebellum — were fulfilled only by 4 of the 22 patients with cognitive impairment of other origin than AD (Fig. 2). The overall specificity of this diagnostic procedure in this sample was 82%. As shown in this study FDG-PET reaches a diagnostic sensitivity and specificity unsurpassed by other imaging modalities (Herholz et al., 1990).

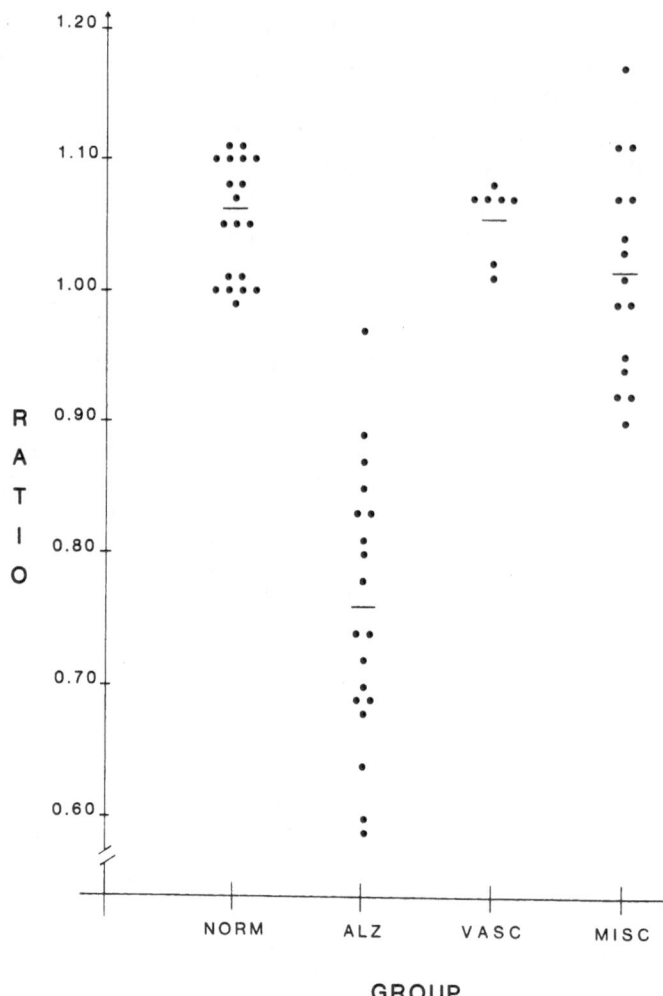

Fig. 2. Scatter plot of the ratio between CMRGlu values in "affected" and "non-affected" brain regions in the various diagnostic groups of: 19 normals, 19 AD patients, 7 patients with vascular dementia and 22 patients with miscellaneous non-Alzheimer diseases. AD is clearly seperated from normals and vascular patients; there is a slight overlap with the miscellaneous group due to cases with dementia in Parkinson disease, posthypoxic and posthypoglycemic states

PET can also be used to assess treatment effects on disturbed metabolism (Heiss et al., 1989). Such studies demonstrated an equalization of metabolic heterogeneities in patients responding to muscarinergic cholinagonists and diffuse increase of metabolism during treatment with piracetam and phosphatidylserine. The therapeutic relevance of such metabolic effects, however, must be proved in controlled clinical trials.

References

Baxter LR, Phelps ME, Mazziotta JC, et al (1987) Local cerebral glucose metabolic rates in obsessive-compulsive disorder — a comparison with rates in unipolar depression and in normal controls. Arch Gen Psychiatry 44: 211–218

DeLeon MJ, Ferris SH, George AE, et al (1983) Computed tomography and positron emission transaxial tomography evaluations of normal aging and Alzheimer's disease. J Cereb Blood Flow Metab 3: 391–394

Duara R, Grady C, Haxby J, et al (1986) Positron emission tomography in Alzheimer's disease. Neurology 36: 879–887

Friedland RP, Budinger TF, Ganz E, et al (1983) Regional cerebral metabolic alterations in dementia of the Alzheimer type: positron emission tomography with (18F)fluorodeoxyglucose. J Comput Assist Tomogr 7: 590–598

Hachinski VC, Iliff LD, Zilkha E, et al (1975) Cerebral blood flow in dementia. Arch Neurol 32: 632–637

Heiss WD, Pawlik G, Herholz K, et al (1984) Regional kinetic constants and CMRGlu in normal human volunteers determined by dynamic positron emission tomography of (18F)-2-fluoro-2-deoxy-D-glucose. J Cereb Blood Flow Metab 4: 212–223

Heiss WD, Herholz K, Pawlik G, et al (1989) Positron emission tomography findings in dementia disorders: contributions to differential diagnosis and objectivizing of therapeutic effects. Keio J Med 38: 111–135

Herholz K, Adams R, Kessler J, et al (1990) Criteria for diagnosis of Alzheimer's disease with positron emission tomography. Dementia 1(3): 156–164

Kuhl DE, Metter EJ, Riege WH, et al (1983) Local cerebral glucose utilization in elderly patients with depression, multiple infarct dementia, and Alzheimer's disease. J Cereb Blood Flow Metab 3 [Suppl 1]: S494–S495

Kuhl DE, Metter EJ, Riege WH, Markham CH (1984) Patterns of cerebral glucose utilization in Parkinson's disease and Huntington's disease. Ann Neurol 15 [Suppl]: S119–S125

Kuhl DE, Metter EJ, Benson DF, et al (1985) Similarities of cerebral glucose metabolism in Alzheimer's and Parkinsonian dementia. J Cereb Blood Flow Metab 5 [Suppl 1]: S169–S170

McKhann G, Folstein M, Katzman R, et al (1984) Clinical diagnosis of Alzheimer's disease. Neurology 34: 939–944

Nahmias C, Garnett ES, Firnau G, Lang A (1985) Striatal dopamine distribution in parkinsonian patients during life. J Neurol Sci 69: 223–230

Szelies B, Karenberg A (1986) Störungen des Glukosestoffwechsels bei Pick'scher Erkrankung. Fortschr Neurol Psychiat 54: 393–397

Authors' address: W.-D. Heiss, M. D., MPI für neurologische Forschung, Klinik für Neurologie, Gleueler Strasse 50, D-W-5000 Köln 41, Federal Republic of Germany

J Neural Transm (1991) [Suppl] 33: 21–26
© by Springer-Verlag 1991

The spectrum of subcortical lesions in MRI, sensitivity and specificity

M. Forsting[1], **W. Hacke**[2], and **K. Sartor**[1]

Departments of [1]Neuroradiology and [2]Neurology, University of Heidelberg,
Federal Republic of Germany

Summary. Subcortical foci of increased signal intensity are frequently identified on MRI in the elderly. The lesions are compatible with various pathologic processes and MRI can only provide supportive data for a suspected diagnosis. Without the patient's clinical history the radiologist is not able to differentiate between real pathologic lesions and physiologic aging processes. The high sensitivity of MRI in detecting white matter lesions and the lack of specificity recommends an excellent teamwork between clinicians and radiologists.

> *Everything can be*
> *everything*
>
> *Anonymus*

Magnetic resonance imaging increasingly used as a diagnostic modality, is highly sensitive to subtle changes of the subcortical brain parenchyma accompany a wide variety of neurologic disorders. While some of these white matter lesions occur in conjunction with known or suspected neurologic disease, others are unexpected or incidental. The question remains whether these latter changes represent early forms of occult neurologic disease or normal physiologic processes. To avoid pathologic or etiologic presumptions, Hachinski et al. (1987) suggested the term leuko-araiosis (LA) to denote areas of decreased attenuation on CT or increased intensity on T2 — weighted MR images.

The goal of this paper is to briefly review the literature regarding the sensitivity and specificity of white matter hyperintensities in T2 — weighted MR images, and to illustrate the problems of interpretation of these lesions in elderly patients.

MRI in vascular disease

As indicated above we are not going to discuss macroangiopathic ischemic lesions, but will instead focus on lesions caused by small vessel occlusion. In 1894 Binswanger (1894) described subcortical arteriosclerotic encephalopathy (SAE) as a microangiopathic leucencephalopathy that was frequently associated with a history of hypertension and clinically characterized by disturbances of memory, mood, and cognition. Additionally one may see focal motor signs and, less often, a pseudobulbar syndrome with deterioration of gait and sphincter control. Before the CT era, postmortem examination was necessary to confirm the diagnosis of SAE. The pathologist found an enlarged ventricular system, demyelination of the centrum semiovale, and cystic lesions in the basal ganglia. Especially small vessels show hypertrophy of the media with fibrosis and hyalinosis. Clinically these patients may present with acute stroke, but more often they show a slowly progressive neurologic and/or neuropsychologic deterioration (Zeumer et al., 1981). CT scans demonstrate ventricular enlargement and bilateral low-attenuation areas in the periventricular white matter (Lotz et al., 1986; Zeumer et al., 1980). On MRI the typical "Binswanger" patient shows a periventricular rim of high signal intensity, sometimes extending into the subcortical U- or arcuate fibers. In some cases, there may be cavities located within this rim, representing areas of white matter infarction, and lacunar infarcts in the basal ganglia or brain stem are common. Recognizing these lesions and knowing the history of hypertension in a specific patient allows one to assume the diagnosis of SAE (Fig. 1).

Currently the diagnosis of SAE is based on clinical criteria and supported by diffuse white matter changes on CT or MRI. An unequivocal diagnosis based on imaging features alone is impossible, but CT and MRI are probably diagnostic if correlated with symptoms and signs and the clinical course.

MRI in degenerative brain disease

There are conflicting reports as to how lesion severity correlates with the clinical diagnosis of dementia. It has been suggested that the two main types of dementia, multiinfarct cognitive disorder (MID) and Alzheimer disease (AD), may be differentiated by the severity of white-matter abnormalities. Bowen et al. (1990) tried to determine the prevalence and severity of white-matter abnormalities detected by MRI in patients with memory disorders and dementia, and to correlate the MR results with clinical data. Their results showed a trend toward a stronger correlation between subcortical lesions and the Hachinski score than between periventricular lesions and this score, suggesting that subcortical lesions may be a more reliable indicator of cerebrovascular disease.

Pathologically, changes in the deep white matter in elderly patients with AD have been characterized by symmetrically distributed partial loss of myelin, axons, and oligodendroglial cells with mild reactive gliosis and fibrosis of arterioles and capillaries, yet without complete white matter infarctions and without signs of hypertensive angiopathy. The pathogenetic basis for the increase in deep white-matter changes in AD is unclear. The study of Gray et al. (1985) on leucencephalopathy associated with amyloid angiopathy may explain the high incidence of white matter MRI lesions in patients with AD. Regardless of the origin, if the pathologic white-matter

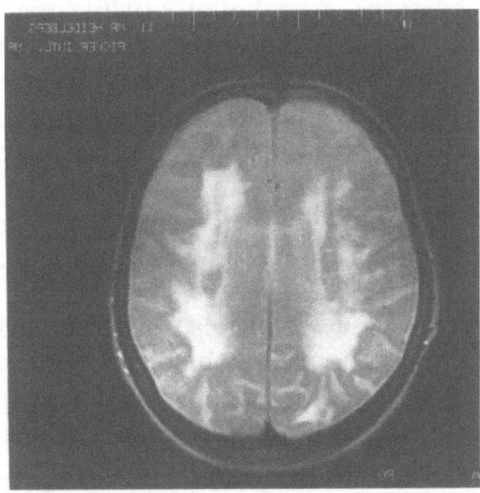

Fig. 1. T2- weighted MRI of a 65 year old female with a history of hypertension and a slowly progressive neurologic deterioration. In conjunction with the patient's history the periventricular rim of hyperintensity allows to assume the diagnosis of SAE

changes in AD cause only an exacerbation of the normal aging process, then MR would show signal abnormalities most prominent in the periventricular watershed zone and relatively less prominent in the subcortical white matter. Bowen et al. (1990) were unable to show a high degree of specificity of MR findings for any subject group with memory disorder or dementia. Nevertheless, the significant differences observed in the mean scores of white-matter lesions in different clinical groups can provide supportive data for a suspected clinical diagnosis. Alternatively, when the expected severity of MR signal abnormalities for a particular clinical diagnosis is not found, other diagnoses should be entertained.

The clinical differentiation between Alzheimer's disease and Binswangers's disease is often difficult. Impaired memory and cognitive functions occur in both, but hypertension, strokes, and asymmetric motor and sensory deficits suggest BD.

MRI in normal elderly persons

Subcortical foci of increased signal intensity are frequently identified on MRI in the elderly, occuring in 30–92% of patients older than 60 years (Awad et al., 1986; George et al., 1986), but data on the clinical significance of these incidental radiological lesions are conflicting. Gerard and Weisberg (1986) reported that these foci are present on MR in the geriatric population with a frequency of 7.8% in asymptomatic patients with no cardiovascualar risk factors; 31% in asymptomatic patients with hypertension, diabetes or heart disease; and 78.5% in patients with cardiovascular risk factors and a history of completed stroke or transient ischemic attack. Roman (1987) argues that these radiologic lesions are indicative of Binswanger's disease. Several postmortem studies revealed that a broad spectrum of pathologic processes underlie the radiologic lesions. Awad et al. (1986) found dilated perivascular spaces, arteriosclerosis and vascular ectasia. Braffman et al. (1988) saw lacunar infarctions, etat crible, white matter infarctions, demyelinative plaques, gliosis, brain cysts and congenital diverticular of the lateral ventricle as the cause of white matter lesions. Other authors found the abnormal white matter characterized pathologically by partial loss of myelin, axons and oligodendroglial cells with mild reactive astrocytic gliosis; additive findings were macrophage infiltration and fibrohyalinization of arterioles (Brun and Englund, 1986).

Awad et al. (1986) proposed that these MRI abnormalities reflect "wear and tear" of the brain parenchyma which accompany aging and chronic cerebrovascular disease. Lesions are usually subcortical, multiple and exhibit increased signal intensity. With the exception of cerebrovascular disease, no significant association with any particular clinical presentation was demonstrated. Age, a prior history of brain ischemia, and hypertension were the most significant predictors of the incidence and severity of white matter lesions.

Like their pathology the pathophysiology of these lesions remains unclear One may find white matter lesions in elderly patients with a history of hypertension and other vascular risk factors but also in patients with a history of *hypo*tension.

DeReuck (1971) proposed a theory stating that the periventricular white matter is a borderzone between the end arteries of the perforating vessels at the base of the skull and descending medullary branches of the main cerebral arteries on the surface of the brain. Therefore, this area is highly susceptible to the effects of decreased blood flow in association with hypotension on one side, and occlusive small vessel disease on the other side. The result will be an incomplete or complete white matter infarction, both resulting in an increased T2 signal on MRI.

The specificity of MRI with regard to these periventricular lesions in older patients is low. Several authors have reported correlations between MRI and neuropathologic findings (Awad et al., 1986; Braffman et al., 1988). In nearly all of these studies the larger lesions were found to represent demyelination, gliosis and areas of infarction. In smaller, more isolated lesions, other pathologic substrates were reported, such as widened perivascular spaces, lacunar infarction, myelin pallor and widened Virchow-Robin spaces. The common factor in all these is probably arteriolar disease. Drayer (1988) proposed that we are looking at a continuum of MRI abnormalities, with Binswanger's disease as the most severe clinical expression and, on the other hand, the "UBOs", or unidentified bright objects, as the first clinical indication of arteriolar change.

Conclusion

Areas of hyperintensity on T2 weighted MR images indicate tissue alterations that are compatible with many pathophysiologic processes. Currently, MRI cannot reliably distinguish between the various pathological processes underlying these white matter lesions. There is no study in the literature that shows a high degree of specifity of MR findings for any subject group with memory disorder or dementia, and it is important to be cautious to consider these lesions as being indicative for dementia or other more specific disorders (Chimonitz et al., 1989; Kobari et al., 1990). White matter lesions seen on MRI can only provide supportive data for a suspected diagnosis. Nearly always clinical or laboratory evidence is necessary to allow a confident diagnosis. The neuroradiologist needs as much information as possible about the patient's history. Aside from SAE, Alzheimer disease and "UBOs" the elderly patients may suffer from multiple sclerosis — late manifestation or late detection — leucencephalopathy after chemo- or radiation therapy, normal pressure hydrocephalus, macroangiopathic hemodynamic lesions, encephalitis, vasculitis etc., all potentially causing similar white matter lesions. Variable pulse sequences and the use of paramagnetics such as Gd-DPTA are possibilities to characterize some of these lesions. The lack of specifity has limited the clinical usefulness of MRI in the presence of unexpected or incidental lesions. In some cases, these incidental findings reveal early or less severe forms of occult neurologic disease — there is a correlation between age and white matter lesions, as well as between hypertension and white matter lesions — but often they are associated with normal physiologic processes. Until now there is much confusion concerning the clinical significance of incidental white matter lesions on MRI, and much remains to be learned about them.

Reading MR images requires a combination of aquaintance with the image formation, analysis of the structural elements, knowledge of disease entities, histopathology, pathophysiology, biochemistry and knowledge of other sources in an assessment that relates to the clinical question (Valk and van der Knaap, 1989).

References

Awad IA, Johnson PC, Spetzler RF, Hodak JA (1986) Incidental subcortical lesions identified on magnetic resonance imaging in the elderly. Part I and II. Stroke 17: 1084–1097

Babikian V, Ropper AH (1987) Binswanger's disease: a review. Stroke 18: 2–12

Binswanger O (1894) Die Abgrenzung der allgemeinen progressiven Paralyse. Berl Klin Wochenschr 31: 1180–1186

Bowen BC, Barker WW, Loewnstein DA, Sheldon J, Duara R (1990) MR signal abnormalities in memory disorder and dementia. AJNR 11: 283–290

Braffman BH, Zimmerman RA, Trojanowski JQ, Gonatas NK, Hickey WF, Schlaepfer WW (1988) Brain MR: pathologic correlation with gross and histopathology. Part I and II. AJNR 9: 621–636

Brun A, Englund E (1986) A white matter disorder in dementia of the Alzheimer type: a pathoanatomical study. Ann Neurol 19: 253–262

Chimowitz MI, Awad IA, Furlan AJ (1989) Periventricular lesions on MRI. Stroke 20: 963–967

DeReuck J (1971) The human periventricular arterial blood supply and the anatomy of cerebral infarction. Eur Neurol 5: 321–334

Drayer PB (1988) Imaging of the aging brain. I. Normal findings. Radiology 166: 785–796

George AE, de Leon MJ, Kalnin A, Rosner L, Goodgold A, Chase N (1986) Leucencephalopathy in normal and pathologic aging. II. MRI of brain lucencies. AJNR 7: 567–570

Gerard G, Weisberg LA (1986) MRI periventricular lesions in adults. Neurology 36: 998–1001

Gray F, Dubas F, Roullet E, Escourolle R (1985) Leucencephalopathy in diffuse hemorrhagic cerebral amyloid angiopathy. Ann Neurol 18: 54–59

Hachinski VC, Potter P, Merskey H (1987) Leuko-araiosis. Arch Neurol 44: 21–23

Kobari M, Meyer JS, Ichijo M, Oravez WT (1990) Leukoaraiosis: correlation of MR and CT findings with blood flow, atrophy and cognition. AJNR 11: 273–281

Lotz PR, Ballinger WE, Quisling RG (1986) Subcortical arteriosclerotic encephalopathy: CT spectrum and pathologic correlation. AJNR 7: 817–822

Miller-Fisher C (1989) Binswanger's encephalopathy: a review. J Neurol 236: 65–79

Roman GC (1987) Senile dementia of Binswanger type: a vascular form of dementia in the elderly. JAMA 258: 1782–1788

Valk J, Knaap van der MS (1989) Magnetic resonance of myelin, myelination and myelin disorders, 1st edn. Springer, Berlin Heidelberg New York Tokyo

Zeumer H, Hacke W, Hündgen R (1981) Subkortikale arteriosklerotische Enzephalopathie. Klinsche, CT-morphologische und elektrophysiologische Befunde. Fortschr Neurol Psychiatr 49: 223–231

Zeumer H, Schonsky B, Sturm KW (1980) Predominant white matter involvement in subcortical arteriosclerotic encephalopathy (Binswanger disease). J Comput Assist Tomogr 4: 14–19

Authors' address: Dr. M. Forsting, Department of Neuroradiology, Im Neuenheimer Feld 400, D-W-6900 Heidelberg, Federal Republic of Germany

J Neural Transm (1991) [Suppl] 33: 27–33
© by Springer-Verlag 1991

Cytoskeleton pathology in Alzheimer's disease and related disorders

F. Seitelberger, H. Lassmann, and **C. Bancher**

Institut für Hirnforschung der Österreichischen Akademie der Wissenschaften and
Neurologisches Institut der Universität Wien, Vienna, Austria

Summary.   The reported findings suggest that ubiquitination of pathological proteinaceous intracytoplasmic inclusions is not at all specific of AD. On the contrary it appears to be a general biochemical marker for disorders in the degradation of a variety of cytoskeletal and other cytoplasmic proteins. The pattern of affected cytoskeletal components is not specific of AD/SDAT tangles. Tau definitely is present also in PSP tangles and possibly in Pick bodies but not in Lewy bodies.

Therefore it has to be considered that the intracytoplasmic accumulation of cytoskeletal protein/ubiquitin complexes in itself is a rather unspecific cellular reaction pattern, possibly a secondary reaction to cell injury of many types, especially, however, of neuronal aging. Nevertheless, the manifestation of NFT in an excessive quantity, intensity, and dynamics with severe concomitant lesions as in AD/SDAT undoubtedly is a true pathological and in this sense a disease-specific change.

Changes of the *neuronal cytoskeleton* are the hallmarks of normal and pathological aging of the brain. This is equally true for brain and aging diseases in the strict sense, resp. their most frequent and severest paragons, e.g. Alzheimer disease and senile dementia of Alzheimer type.

In this presentation recent studies into the nature of *cyto-skeleton neuropathologies* by means of immunocytochemical methods are reported.

Conventional *neuropathology of Alzheimer's disease* comprises degenerative-rareficient as well as productive-accumulative changes. Characters of the latter are: *neurofibrillary tangles* and *senile plaques*. *Neurofibrillary tangles* (NFT) are intracellular accumulations of fibrillary material, which can be visualized by silver impregnation techniques. Ultrastructurally the tangles are composed of pathological filaments, either in the form of "*paired helical filaments*" (PHF; diameter 22 µm) and/or of *straight tubules* (diameter 15 µm). Regarding size and intrinsic composition different *stages* of neurofibrillary tangle formation can be found (Alzheimer, 1911). Early tangles (stage 1) are fine fibrillar or rod-shaped argyrophilic inclusions, mainly located in paranuclear position. Mature tangles (stage 2) fill out most of the cytoplasm of involved nerve cells. In pyramidal cells they are mainly flame-shaped. After neuronal cell death the fibrillar structures become freely located in the extracellular space: these *"ghost"* *tangles* (stage 3) still indicate the original shape of the neurons. Cell nuclei are not found with ghost tangles and their argyrophilic fibrils are intermingled with ingrowing glial cell processes.

On the contrary, *senile plaques* are extracellular deposits and mainly composed of *amyloid fibrils* (diameter 5–10 µm). They form the amyloid plaque core which is in variable extent surrounded by dystrophic nerve cell processes (mostly neurites). These dystrophic neuronal processes frequently contain PHF.

Neurofibrillary tangles (NFT): Most immunocytochemical studies show an incorporation of epitopes of *cytoskeletal proteins* into the fibrillar material of NFT.

Numerous studies documented reactivity of NFT to antisera or monoclonal antibodies against *phosphorylated neurofilaments* (NF) (Anderton et al., 1982). In normal conditions phosphorylated neurofilament epitopes are present in axons, whereas non-phosphorylated determinants are found in pericarya and dendrites (Sternberger et al., 1985). This difference seems to be connected with the intraneural neurofibril production and transport. Furthermore, phosphorylation of neurofilaments is probably required for the stabilisation (protection against proteolysis) of the axonal cytoskeleton. Interestingly, Alzheimer NFT selectively react to some antibodies against phosphorylated epitopes (Sternberger et al., 1985).

In recent studies it turned out that the neurofilament reactivity of NFT is due to shared antigenic determinants between certain phosphorylated neurofilament epitopes and phosphorylated epitopes of the *tau protein*. This cytoskeletal component: the low molecular weight *microtubule associated protein tau* (MAP tau; 55–70 kD) consistently is present in NFT (Brion et al., 1985). Recent studies indicate that tau proteins are incorporated into tangles in a pathologically phosphorylated modification (Grundke-Iqbal et al., 1986b; Flament et al., 1989 etc.)

Another series of studies indicates the presence of antigenic determinants which seem to be *unique to NFT*. These observations are based on the reactivity of tangles to poly- or monoclonal antibodies, which do not recognize structures in normal CNS tissue (Ihara et al., 1983; Wang et al., 1984 etc.). At present it became obvious that these antibodies recognize determinants of tau protein, of not identified components of NFT (possibly true neoantigens) and also of *ubiquitin*. Ubiquitin is a small polypeptide (4500 D), which is involved in the non-lysosomal protein breakdown. Apparently proteins are bound to ubiquitin and then the ubiquitin/protein complexes are degraded by peptidases. Following protein degradation ubiquitin is released and can be reutilized in the same type degradation processes.

The molecular and pathogenic relationship between NFT and *plaque amyloid* is still an open question. Plaque amyloid is formed by aggregates of fibrils consisting of the so-called amyloid beta-peptide (A4 polypeptide). Similar to plaque amyloid also PHF of NFT are beta-pleated sheaths and thus stained by conventional amyloid dyes. However, a direct relationship between plaque amyloid and tangles cannot be stated. Furthermore the guessed common genetic source of amyloid and NFT appears not supported by molecular genetic findings.

During the last years our group performed immunocytochemical studies in the composition of NFT starting from two special questions:

1) Which antigens are present in tangles at different stages of tangle formation and maturation?

2) What antigenic profile is unique to Alzheimer NFT?

ad 1) For the study of *NFT immunochemistry* we used a spectrum of poly- and monoclonal antibodies directed against NF subunits, tubulin, MAP1, MAP tau, and ubiquitin.

Overall we found selective staining of tangles with antibodies against PHF, against some phosphorylated NF epitopes, against tau and ubiquitin. By ultrastructural immunocytochemistry tau and ubiquitin reactivity was associated to the PHF.

On the contrary immune reactivity for non-phosphorylated NF epitopes as well as for MAP1, MAP2 and tubulin was diffusely distributed throughout the neuronal pericaryon, without selective labelling of NFT.

Developmental stages of NFT formation: The study in quantitative evaluation of the immunoreactivity of tangles revealed that about equal numbers of tangles were stained by Bielschowsky's silver stain and by immunocytochemistry with a polyclonal anti-PHF serum. However, in adjacent sections the number of immune reactive tangles reactive to anti-tau or anti-ubiquitin antibodies was substantially lower. This made us suspect that the antigenic profile of NFT might change with maturation of these fibrillar inclusions. We thus correlated the immunoreactivity of the tangles with the *maturation stages* described by Alzheimer (1911) (Bancher et al., 1989a). The results were:

NFT of all stages strongly reacted to *polyclonal anti-tau* antibodies whereas *reactivity* with the *tau 1 monoclonal antibody* was found mainly in small resp. early (stage 1) tangles and in classical flame shaped (stage 2) tangles. Extracellular "ghost" tangles (stage 3), however, in general were tau 1 negative.

On the contrary *ubiquitin* was mainly accessible in stage 2 and in "ghost" tangles, which were also positive for GFAP (due to the ingrowth of glial cell processes between the extracellularily located pathological fibrillar bundles). *Ubiquitin* reactivity was generally absent in early stage 1-tangles.

In addition in cases of Alzheimer's disease we found a neuronal population, especially in the hippocampus, which in the cytoplasm showed diffuse reactivity to anti-tau antibodies similar to that described for Alz 50 (Hyman et al., 1987). By silver stain these neurones did not exhibit NFT. Ultrastructurally tau reactivity in such neurons was associated to single PHF. These data thus suggest, that immunostaining with tau antibodies renders the recognition of a very early stage of tangle formation, which precedes the morphological appearence of argyrophilic fibrillar inclusions. This pattern of NFT immune reactivity we named "*stage 0*" tangles.

These data could either mean that with maturation of tangles more and more *ubiquitin is incorporated* into the fibrils (ubiquitination), which finally masks the tau epitopes. Alternatively and more probably, primary tau-ubiquitin complexes in the tangles may undergo conformational changes or partial degradation, resulting in inaccessibility of tau epitopes and exposure of the previously hidden (3–39) ubiquitin epitopes.

Double immunostaining with anti-tau and anti-MAP 1, anti-MAP 2, anti tubulin or anti-NF respectively were performed also. Although tangle bearing neurons frequently showed irregular, distorted dendrites, the pericaryal and dendritic reactivity of these neurons for MAP's or tubulin was not different from that of normal neurons. However, neurons with early (especially stage 1) tangles frequently revealed cytoplasmic reactivity in the pericaryon and dendrites with a monoclonal antibody against a phosphorylated epitope of neurofilaments (SMI 31). That seems to point to a more complex disorder of the cytoskeleton from which the conditions for production of NFT as a not strictly specific type of neuronal reaction originate.

In *summary* the results of our examination of *Alzheimer NFT* indicate that pathologically phosphorylated tau epitopes appear at the earliest stages of NFT formation. With tangle maturation certain ubiquitin epitopes become accessible for antibodies in the pathological fibrils, suggesting either progressive ubiquitination or conformational changes of PHF during tangle maturation. At present it is unresolved, whether additional protein antigens are present in PHF and to what extent glycosaminoglykanes are incorporated in various forms of tangles. Another interesting feature is the stronger staining of ghost tangles with amyloid dyes and positive β-peptide reactivity: That may indicate that the modification of the extracellular NFT material could become the nidus for β-peptide amyloid incorporation and possibly also formation.

ad 2) Although these cytoskeletal abnormalities are highly characteristic for Alzheimer's disease, the question arises, to what extent these alterations are unique for AD/SDAT. We therefore investigated by using the same type of technology several *other cytoskeleton disorders* of the central nervous system, characterized by neuronal or glial cytoskeleton abnormalities, respectively intracellular protein inclusions, among others Progressive supranuclear palsy, Pick's disease, and Parkinson's disease.

Progressive supranuclear palsy (PSP): is a neurodegenerative disease clinically characterized by dementia, hypertonus of neck and shoulder girdle muscles and in some cases Parkinsonian symptoms. The pathognomonic change of PSP are NFT in the neurons of subcortical grisea, esp. substantia nigra, locus coeruleus, nucleus basalis, subthalamic nucleus, and substantia innominata. These inclusions are mostly of scein- or ball-like form and in conventional studies very similar to Alzheimer NFT. Ultrastructurally these NFT are mainly composed of 15μm straight tubules with only few or absent PHF. Overall, PSP tangles have a similar antigenic profile compared to AD/SDAT tangles. Particularily, similar as in AD/SDAT NFT, the inclusions in PSP strongly react to antibodies against tau. Although the presence of some ubiquitin epitopes has been described in PSP tangles (Manetto et al., 1988), the 3–39 ubiquitin epitope, an antigenic determinant which appears in late stages of tangle maturation in AD/SDAT, is absent in PSP tangles. PSP tangles, thus, reveal an antigenic profile, similar to early (immature ?) tangles in AD/SDAT (Bancher et al., 1987).

Pick's disease: This is another example of presenile dementia. In contrary to Alzheimer disease it is characterized by disintegration of personality with loss of intentional drive and of selfconcern. Neuropathology is dominated by severe neuronal loss in circumscribed areas of the cerebral cortex. In affected neurons NFT particular in structure and topographical distribution can be found, especially in the pathognomonic form of the argyrophilic *Pick bodies*. Ultrastructurally the Pick bodies contain a mixture of filamentous elements of different structure, including normal NF, as well as 15µm straight tubules and some PHF. Thus it is not surprising that anti-PHF sera also stain Pick bodies in variable degrees (Rasool et al., 1985). More specifically epitopes of phosphorylated neurofilaments as well as of ubiquitin have been detected in Pick bodies (Manetto et al., 1988; Munoz Garcia and Ludwin, 1984; Rasool et al., 1985; Yen et al., 1986).

Parkinson's disease: is a frequent neurodegenerative disorder characterized by muscle hypertonus and tremor. Neuropathology shows selective loss of pigmented neurons of substantia nigra and locus coeruleus. Due to the nigra-lesions the concentration of dopamine in the neostriatum is highly reduced. A characteristic histological feature is the appearance of *Lewy bodies* in the cytoplasm of affected neurons. Lewy bodies are found in widespread distribution in diffuse Lewy body disease (Kosaka et al., 1981) which exhibits a Parkinsonian syndrome complicated by dementia. Lewy bodies are round, hyaline-like inclusions, ultrastructurally composed of a core of amorphous, granular material with some membraneous, sometimes vesicular material, surrounded by a halo of radiating fibrils, which structurally resemble neurofilaments. Immunocytochemically Lewy bodies contain epitopes of neurofilaments and ubiquitin. NF determinants are mainly present in the peripheral halo of Lewy bodies. In untreated paraffin sections reactivity is found only with antibodies against phosphorylated NF epitopes. However, by predigestion of the sections with alkaline phosphatase, non-phosphorylated NF epitopes can be unmasked. Thus, neurofilament/ubiquitin complexes appear to be a major component of Lewy bodies. Contrary to AD/SDAT tangles Lewy bodies do not react with anti-tau antibodies (Bancher et al., 1989b).

Rosenthal fibers: Rosenthal fibers are hyaline inclusions in astrocytes, which can be found in a variety of gliomas and in especially high numbers in astrocytes of *Alexander's disease*. Alexander's disease is a neurodegenerative condition mostly manifesting in early age by neuropsychiatric retardation. Neuropathology is characterized by a large heavy brain and diffuse cerebral white matter destruction. Ultrastructurally Rosenthal fibers have a central amorphous core, surrounded by a periphery composed of typical intermediate filaments. Immunocytochemically Rosenthal fibers react with antibodies against GFAP and ubiquitin, but do not contain epitopes of MAP1, MAP2 or tau or of NF. It is thus suggested that GFAP/ubiquitin complexes are a major component of Rosenthal fibers.

Other inclusions: In *Marinesco bodies* ubiquitin can be found, although it does, however, not appear to be associated with cytoskeletal proteins. Interestingly, also *Corpora amylacea*, aging products of glial cells, contain ubiquitin. It is not established to what other protein(s) ubiquitin is conjugated in these inclusions.

References

Alzheimer A (1907) Über eine eigenartige Erkrankung der Hirnrinde. Allg Z Psychiatr 64: 146

Alzheimer A (1911) Über eigenartige Krankheitsfälle des späteren Alters. Z Ges Neurol Psych 4: 356–385

Anderton BH, Breinburg D, Downes MJ, Green PJ, Tomlinson BE, Ulrich J, Wood JN, Kahn J (1982) Monoclonal antibodies show that neurofibrillary tangles and neurofilaments share antigenic determinants. Nature 298: 84–86

Bancher C, Lassmann H, Budka H, Grundke-Iqbal I, Iqbal K, Wiche G, Seitelberger F, Wisniewski HM (1987) Neurofibrillary tangles in Alzheimer's disease and progressive supranuclear palsy: antigenic similarities and differences. Acta Neuropathol (Berl) 74: 39–46

Bancher C, Brunner C, Lassmann H, Budka H, Jellinger K, Wiche G, Seitelberger F, Grundke Iqbal I, Iqbal K, Wisniewski HM (1989a) Accumulation of abnormally phosphorylated tau precedes the formation of neurofibrillary tangles in Alzheimer's disease. Brain Res 477: 90–99

Bancher C, Lassmann H, Budka H, Jellinger K, Grundke Iqbal I, Iqbal K, Wiche G, Seitelberger F, Wisniewski HM (1989b) An antigenic profile of Lewy bodies: immunocytochemical indication for protein phosphorylation and ubiquitination. J Neuropathol Exp Neurol 48: 81–93

Brion JP, Passareivo H, Nunez J, Flament-Durand J (1985) Mise en évidence immunologique de la proteine tau au niveau des lésions de degenerescence neurofibrillaire de la maladie d'Alzheimer. Arch Biol 95: 229–235

Flament S, Delacourte A, Hemon B, Defossez A (1989) Characterization of two pathologcal tau protein variants in Alzheimer brain cortices. J Neurol Sci 92: 133–141

Gajdusek DC (1985) Hypothesis: interference with axonal transport of neurofilament as a common pathogenetic mechanism in certain diseases of the central nervous system. N Engl J Med 312: 714–717

Grundke-Iqbal I, Johnson AB, Wisniewski HM, Terry RD, Iqbal K (1979) Evidence that Alzheimer neurofibrillary tangles originate from neurotubules. Lancet i: 578–580

Grundke-Iqbal I, Iqbal K, Quinlan M, Tung YC, Zaidi MS, Wisniewski HM (1986a) Microtubule-associated protein tau. A component of Alzheimer paired helical filaments. J Biol Chem 261: 6084–6089

Grundke-Iqbal I, Iqbal K, Tung YC, Quinlan M, Wisniewski HM, Binder LI (1986b) Abnormal phosphorylation of the microtubule-associated protein T (tau) in Alzheimer cytoskeletal pathology. Proc Natl Acad Sci USA 83: 4913–4917

Hyman BT, van Hoesen GW, Wolozin B, Davis P, Kromer LJ, Damasio AR (1987) Alz 50 antibody recognizes Alzheimer-related neuronal changes. Ann Neurol 23: 371–379

Ihara Y, Nukina N, Miura R, Ogawara M (1986) Phosphorylated tau protein is integrated into paired helical filaments in Alzheimer's disease. J Biochem 99: 1807–1810

Kosaka K, Matsushita M, Oyanagi S, Mehraein P (1980) A clinico-neuropathological study of the "Lewy body disease". Seishin Shinkeigaku Zasshi 82: 292–311

Manetto V, Gambetti P, Tabaton M, Mulvihill P, Fried V, Smith H, Autilio-Gambetti L, Perry G (1987) Ubiquitin conjugates: a new component of abnormal neuronal filaments in neurodegenerative diseases. Soc Neurosci Abstr 13: 820

Munoz-Garcia D, Ludwin SK (1984) Classic and generalized variants of Pick's disease: a clinicopathological, ultrastructural, and immunocytochemical comparative study. Ann Neurol 16 (4): 467–481

Rasool CG, Abraham C, Anderton BH, Haugh M, Kahn J, Selkoe DJ (1984) Alzheimer's disease: immunoreactivity of neurofibrillary tangles with anti-neurofilament and anti-paired helical filament antibodies. Brain Res 310: 249–260

Sternberger NH, Sternberger LA, Ulrich J (1985) Aberrant neurofilament phosphorylation in Alzheimer disease. Proc Natl Acad Sci USA 82: 4274–4276

Yen SH, Horoupian DS, Terry RD (1983) Immunocytochemical comparison of neurofibrillary tangles in senile dementia of Alzheimer type, progressive supranuclear palsy, and postencephalitic parkinsonism. Ann Neurol 13: 172–175

Authors' address: em. o. Prof. Dr. F. Seitelberger, Neurologisches Institut der Universität, Schwarzspanierstrasse 17, A-1090 Wien, Austria

J Neural Transm (1991) [Suppl] 33: 35–38
© by Springer-Verlag 1991

Morphometry of the corpus callosum in normal aging and Alzheimer's disease

S. Weis[1], **K. Jellinger**[2], and **E. Wenger**[3]

[1]Institute of Anatomy, Medical University Lübeck, Federal Republic of Germany
[2]L. Boltzmann Institute of Clinical Neurobiology, Vienna, and [3]Institute for Information Processing, Austrian Academy of Sciences, Vienna, Austria

Summary. Changes of the human corpus callosum in normal aging and Alzheimer's disease were analysed by means of morphometry. A standardized computerized evaluation program was implemented allowing objective, quantitative and reproducible data. The various parts of the corpus callosum showed a different pattern of changes in normal aging as compared to Alzheimer's disease. In conclusion, in normal aging affects mostly the front-temporal interhemispheric fiber systems, whereas in Alzheimer's disease the parietotemporal commissural fibers are altered.

The corpus callosum plays an important role in the interhemispheric information transfer. Little is known about its involvement in normal and pathological aging. The aim of the present study was to determine by means of morphometry the importance of aging processes in the human corpus callosum.

Median-sagittal planes from the brains of 48 normal persons obtained by nuclear magnetic resonce imaging (NMR) were analysed. The sample of the autopsy brains consisted of 9 control and 9 Alzheimer's brains.

Since the variability of the shape of the corpus callosum is very high, we first implemented a systematic quantification strategy which can easily be reproduced for further comparative studies.

The quantification was performed in the following way: The images were entered by a video camera into the computer system. The outline of the corpus callosum was traced with the digitizer. The metric scale as well as the lowest points of rostrum and splenium were transmitted to the system. The inferior points of rostrum and splenium were connected by a tangent, called horizontal inferior 1 (HI-1). A vertical line, called vertical anterior 1 (VA-1), was traced perpendicular to the horizontal inferior 1 and tangent to the most anterior point of the genu.

A vertical line, called vertical posterior 1 (VP-1), was traced perpendicular to the horizontal inferior 1 and tangent to the most posterior point of the splenium. A horizontal line, called horizontal superior 1 (HS-1), was traced perpendicular to the vertical anterior 1 and vertical posterior 1, parallel to

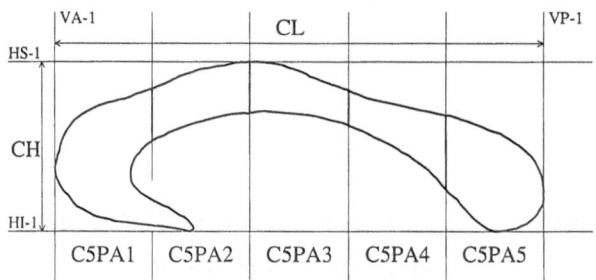

Fig. 1. Morphometric evaluation of the human corpus callosum (for details and abbreviations see text)

the horizontal inferior 1 and tangent to the most superior point of the trunk (Fig. 1). For further methodological details see Weis et al. (1989).

The result of this procedure was a rectangular figure which enabled us to measure a) callosal height (CH) as the distance between HI-1 and HS-1 and b) callosal length (CL) as the distance between VA-1 and VP-1 (Fig. 1).

The callosal length was divided into two, three, four and five parts by lines which were traced perpendicular to the HI-1 and HS-1 (Fig. 1). The profile area of the different parts of the corpus callosum defined in this way were consequently evaluated.

Analysis of the NMR images showed that in normal aging a significant decrease of profile area occurs in the most anterior parts (rostrum, genu, and anterior parts of trunk) of the corpus callosum, whereas the posterior parts (posterior parts of trunk and splenium) remain unchanged (Table 1). The same pattern was found in autopsy brains (Table 2).

In Alzheimer's disease a significant decrease occurs in the middle parts (trunk) of the corpus callosum, whereas the most anterior and posterior parts

Table 1. Results of the morphometric evaluation of the human corpus callosum in normal aging. The quantification was performed on mid-sagittal NMR images

	ACL 1 (n = 19) (20 − 39) MA = 30.3		ACL 2 (n = 16) (40 − 59) MA = 48.6		ACL 3 (n = 13) (60 − 80) MA = 70.3		F ratio	p
	MW	SD	MW	SD	MW	SD		
CPA	697.00	(85.4)	620.59	(75.9)	612.31	(139.2)	3.71	0.03
CL	74.71	(4.9)	71.43	(4.5)	72.58	(5.9)	1.85	0.17
CH	26.50	(2.9)	23.83	(1.8)	26.62	(3.0)	5.46	0.008
C5PA1	211.80	(27.9)	179.15	(29.1)	179.84	(43.9)	5.41	0.008
C5PA2	105.02	(18.2)	93.21	(15.8)	86.27	(20.9)	4.37	0.02
C5PA3	93.74	(14.4)	83.22	(9.9)	84.15	(24.3)	2.14	0.13
C5PA4	91.73	(22.4)	87.67	(16.6)	78.43	(25.5)	1.48	0.24
C5PA5	194.70	(22.7)	177.34	(25.2)	183.62	(39.1)	1.64	0.21

Table 2. Results of the morphometric evaluation of the human corpus callosum in normal aging and in Alzheimer's disease. The quantification was performed on autopsy brains

	ACL 1 (n = 19) mean age = 45.8		ACL 2 (n = 21) mean age = 82.4		Mann-Whitney-test
	mean	sd	mean	sd	p
CPA	841.83	(166.3)	636.32	(88.7)	0.000
CL	83.58	(8.9)	83.42	(6.1)	0.80
CH	24.24	(3.9)	20.37	(2.6)	0.002
C5PA1	285.53	(66.9)	205.79	(31.9)	0.000
C5PA2	124.67	(27.4)	90.91	(17.3)	0.000
C5PA3	116.38	(24.6)	88.13	(13.2)	0.000
C5PA4	126.38	(28.9)	96.03	(20.8)	0.001
C5PA5	188.85	(41.7)	155.45	(34.1)	0.01

	Control (n = 9) mean age = 72.5		Alzheimer (n = 9) mean age = 79.1		Mann-Whitney-test
	mean	sd	mean	sd	p
CPA	686.47	(129.9)	591.91	(73.0)	0.06
CL	76.18	(4.6)	78.15	(3.7)	0.31
CH	20.62	(3.2)	19.27	(3.2)	0.33
C5PA1	200.03	(32.3)	167.86	(46.2)	0.10
C5PA2	100.65	(22.0)	83.21	(10.6)	0.08
C5PA3	102.79	(26.9)	82.94	(14.9)	0.05
C5PA4	123.88	(34.3)	105.23	(22.5)	0.17
C5PA5	159.10	(27.4)	152.66	(20.2)	0.76

remain unchanged (Table 2). Besides these two-dimensional parameters, further evaluations of one-dimensional parameters on callosal thickness and width have been carried out. These data showed the same trend and confirmed the above results.

In summary one can say that normal aging alters mostly the frontal and temporal interhemispheric fiber systems. In Alzheimer's disease the parietal and temporal commissural fibers are strongly affected, whereas a certain amount of frontal and occipital interhemispheric fibers are less affected.

The results are in accordance with other morphometric papers that reported a decrease of the frontal cortex in normal aging, whereas the salient changes in Alzheimer's disease occur in the parietal and temporal cortex (Terry, 1986).

The next step in our research on the involvement of the corpus callosum in normal and pathological aging will be directed towards the quantification of the axons. Further investigations will be carried out (1) to correlate callosal data with other morphometric data of cerebral structures derived from horizontal NMR imaging planes and (2) to analyze more reliably the involvement of the corpus callosum in normal and pathological aging with neuropsychological tests.

References

Terry RD (1986) Interrelations among the lesions of normal and abnormal aging of the brain. In: Swaab DF, Fliers E, Mirmiran M, Van Gool WA, Van Haaren F (eds) Aging of the brain and Alzheimer's disease. Elsevier, Amsterdam, pp 41–48 (Prog Brain Res 70)

Weis S, Weber G, Wenger E, Kimbacher M (1989) The controversy about a sexual dimorphism of the human corpus callosum. Int J Neurosci 47: 169–173

Authors' address: Dr. S. Weis, Institute of Neuropathology, University Munich, Thalkirchnerstraße 36, D-W-8000 Munich 2, Federal Republic of Germany

J Neural Transm (1991) [Suppl] 33: 39–48
© by Springer-Verlag 1991

Clinical and epidemiological aspects of dementia in the elderly

P. Fischer[1,2] and **P. Berner**[2]

[1]Neurological Institute, University of Vienna, and [2]Psychiatric Clinic, Vienna Medical School, Vienna, Austria

Summary.   Dementia of the Alzheimer's type (DAT) is the most significant disease of the aging brain. Descriptive epidemiology of DAT found a constant doubling of prevalence rates every 5 years. Analytic epidemiology so far failed to reliably detect risk factors for DAT other than age. This might depend on the difficulties encountered in the clinical diagnosis and differential diagnosis of dementia in the elderly, which are discussed with special reference to 1) the definition of dementia, to 2) the grading of severity of dementia, to 3) the differentiation between dementia and depression, and to 4) the differentiation between multi-infarct dementia and DAT.

Descriptive epidemiology of dementia in the elderly

At least 2% of those aged 70 to 79 years and at least 5% of those older than 80 years need chronic care because of an advanced dementing illness. Prevalence rates of dementia including mild and moderate stages are many times higher (Mölsä et al., 1982; Kay, 1986; Jorm et al., 1987; Cooper and Bickell, 1989; Evans et al., 1989). Jorm et al. (1987), metaanalysing the results of 47 epidemiological studies on dementia, found a constant doubling of prevalence rates every 5.1 years. Age specific prevalence rate for dementia is 0.7% in the 60–64 years old, increases over 5% between 75 and 79 years (5.6%) and is 38.6% in those aged 90–95 years. Progressive changes in the age distribution of the population due to increased life-expectancy will lead to a veritable epidemic of dementia in the future. On the other hand, a mild decrease of incidences of dementia of the Alzheimer's type (DAT) and multi-infarct dementia (MID) in the past 30 years has been described in two studies (Hagnell et al., 1983; Kokmen, 1989).

At autopsy more than 60% of dementias in the elderly are diagnosed as DAT (Alzheimer, 1907; Fischer, 1907; Simchowicz, 1911; Tomlinson et al., 1970; Alafuzoff et al., 1987; Jellinger, 1976; Jellinger et al., 1990). In about one-third of these cases cortical and subcortical Lewy bodies are found together with Alzheimer's pathology (Perry et al., 1990; Hansen et al., 1990). The significance of this finding for cognitive and behavioral symptoms needs further evaluation. No differences have been found either on the clinical-neuropsychological level or on the neuropathological-neuro-immunocytochemical level between presenile and senile DAT (Amaducci et al., 1986). About 20% of dementias in the elderly are caused by multiple cerebral infarcts (MID); another 10% result from a combination of DAT and MID called mixed dementia (MIX). More than 60 other dementing conditions must be discriminated from MID and DAT (Marsden, 1985; Katzmann, 1986).

Both, DAT and MID, are age-dependent disorders (Katzmann, 1988): the incidences of both diseases increase exponentially with age. The question whether the incidence of DAT decreases in the very old, as found in Parkinson's disease (Schoenberg, 1986), has not yet been answered. Clarification would give important information whether DAT is the exaggeration of normal aging (Miller et al., 1984) or whether it is an age-dependent disorder (Berg, 1985).

Analytic epidemiology of dementias in the elderly

The risk factors for cerebrovascular accidents are well known and extensively discussed in the literature (Hachinski, 1985; Tuszynski et al., 1989). The only unequivocally accepted risk factors for DAT are a patient's age and a positive family history of DAT. Various other risk factors for DAT, including head trauma, thyroid disease, family history of Down's syndrome, and maternal age, have been proposed and rejected in following studies (Henderson, 1986, 1988; Kay, 1986; Rocca et al., 1986; Fitch et al., 1988; Chandra et al., 1989; Farrer et al., 1989). Recent studies suggest lower rates for DAT in both, China and Japan compared to western countries (Jorm et al., 1987; Henderson, 1988; Shibayama et al., 1986; Ng and Lee, 1988; Li et al., 1989), which is open to explanation on genetic or environmental grounds.

It has been found that DAT is about 50% more frequent in women than in men (Mölsä et al., 1982; Jorm et al., 1987; Henderson, 1988). Up to now no neuropathological investigation found differences regarding intensity or quality of lesions of the Alzheimer type between female and male patients (Miller et al., 1984). We speculate that dementia in women is overdiagnosed: depression is more frequently found in elderly women (Hagnell et al., 1982), and thus, depressive pseudodementia might also have different prevalences in both sexes; moreover, demented women might live longer than demented men, which would cause different prevalence but not incidence rates for DAT (Knopman et al., 1988; Kokmen, 1989).

The definition of dementia

Eugen Bleuler (1916) defined dementia as the most severe end-stage of the organic brain syndrome, a definition which included severity and irreversibility of cognitive decline and largely influenced the concept of dementia and organic brain syndrome (ICD-9, 1971; Berner, 1977). In 1980 the American

Psychiatric Association redefined dementia. This concept of an organic cognitive impairment, not diagnosed exclusively during delirium, including at least memory and one more cognitive domain, and being severe enough to influence social or occupational functioning, is now used worldwide. The new definition of dementia in the ICD-10 will also approach the definition of the DSM-III-R (1987). The remaining problem regarding the definition of dementia in clinical praxis is how to define "mild" dementia and how to discriminate it from benign senescent forgetfulness (Kral, 1962), also called age-associated memory impaiment (AAMI, Crook et al., 1986), and other cognitive syndroms found by elaborate neuropsychological testing.

In DSM-III-R (1987) severity of dementia is defined by the severity of impairment of daily living (work, social activities, personal hygiene, . . .) and not by the severity of the underlying cognitive impairment. Therefore, MID patients might be rated as more severely demented when compared to patients with other dementing disorders, because a MID patient may be incapable of self-care owing to his neurological deficits as well as to his cognitive impairment. Psychosocial parameters, such as social class, living alone, distance and complexity of the route to shops, etc., also influence the diagnosis of dementia by this definition. Patients with more intellectually demanding professions like lawyers, teachers, etc., will be detected earlier in the course of their dementing illness than patients working in less demanding jobs. This might explain why epidemiological studies found lower rates of dementia in rural scandinavian communities (Jorm et al., 1987).

Another possibility to grade severity of dementia is by the help of commonly accepted short dementia rating scales like the Mini-Mental State Examination MMSE or the Blessed Orientation-Memory-Concentration Scale BOMCS (Blessed et al., 1968; Folstein et al., 1975; Folstein, 1983; Thal et al., 1986; Fillenbaum et al., 1987; Bleeker et al., 1988; Galasko et al., 1990). The problem with these scales is that they are not objective because instructions for execution and scoring and for interpretation of test scores are lacking. Moreover, the decision to call a patient demented cannot be made by a simple short rating scale, but is the result of a long and time-involving diagnostic process including two major differentiations: firstly, the discrimination of normal aging including AAMI from mild dementia and secondly, the correct discrimination between depression with cognitive impairment ("depressive pseudodementia") and dementia with depressive symptomatology ("demented pseudodepression").

Dementia and/or depression

The incidence of depression increases with increasing age (Hagnell et al., 1982). Depression is diagnosed about four times more frequently than dementia in the elderly (Kay et al., 1985; Siegel and Gershon, 1987; Kivelä et al., 1988). Therefore, the discrimination between dementia and pure depression is the most frequent question raised in gerontopsychiatry (Kiloh, 1961; Wells, 1979; Good, 1981; Caine, 1981; Kral, 1982; Rabins et al., 1984;

Siegel and Gershon, 1987; Pearlson et al., 1989; Danielczyk et al., 1989). About 10% of depressions in the elderly present clinically as dementia syndrome (Gurland and Toner, 1987), a phenomenon which has been labelled "depressive pseudodementia" (Madden et al., 1952). Moreover, there are cognitive impairments (concentration, memory, speed of thinking) in nearly every depressed patient (Strömgren, 1977; Weingartner et al., 1981). Old depressed patients frequently complain about memory problems, but these complaints are not correlated with cognitive impairment; they are rather correlated with severity of depression (Pettinati et al., 1985).

Both depression and depression-like symptomatology occur frequently in the course of MID and DAT (Reifler et al., 1982; McAllister and Price, 1982; Knesevich et al., 1983; Cummings et al., 1987; Reisberg et al., 1987; Lazarus et al., 1987; Zubenko and Moossy, 1988; Venna et al., 1988; Wragg and Jeste, 1989). While depression decreases in the course of DAT, depressive symptoms increase in MID in relation to physical disability and helplessness (Danielczyk et al., 1989; Fischer et al., 1990c). Taken together there is a great need for longitudinal studies concerning the clinical differentiation of (mild) dementia and/or depression (Roth, 1955).

A problem in investigating depression in the elderly demented patient is that standardized depression scales for this population are not available (Weiss et al., 1986). Two recently developed scales need further validation (Brink et al., 1982; Yesavage and Brink, 1983; Sheikh and Yesavage, 1986; Sunderland et al., 1988).

The differentiation between MID and DAT

The diagnostic evaluation of a questionable demented patient has to be carried out in 3 steps. The first step is to decide whether the patient is really demented: delirium, focal neurological deficit and pseudodementia due to depression or schizophrenia has to be excluded (Marsden, 1985). While a clearly abnormal EEG is an important hint to a delirium, a normal EEG should raise suspicion of pseudodementia (Marsden, 1985; Streifler et al., 1990). Then follows the exclusion of dementing conditions other than MID or DAT and the exclusion of treatable dementias, especially (hypothyreosis, vitamine B_{12}-deficiency, chronic Wernicke-encephalopathy, frontal meningioma, progressive paralysis, . . .). In most cases the third step will be the differentiation between dementia of vascular origin and primary degenerative dementia. This differentiation is possible in about two-thirds of demented patients (Sulkava et al., 1983; Mölsä et al., 1985; Alafuzoff et al., 1987; Erkinjuntti et al., 1988; Joachim et al., 1988; Tierney et al., 1988; Boller et al., 1989; Jellinger et al., 1990). Probability-corrected classification by chance would also diagnose 2/3 of the patients (Meehl and Rosen, 1955). Because an intra-vitam diagnostic test is still lacking, diagnostic validity is only possible in selected samples when the rather high percentage of atypical patients is excluded from study (Fischer et al., 1990a).

The application of common clinical criteria (DSM-III-R, 1987; NINCDS-ADRDA: McKhann et al., 1984) correctly classifies about 80% of DAT and about 60% of MID patients (Morris et al., 1988; Tierney et al., 1988). A possible cause of the more difficult diagnosis of MID is the overinterpretation of CT, EEG and NMR (Ettlin et al., 1989). But even course characteristics mislead in the differentiation of MID and DAT (Fischer et al., 1989, 1990a). While most DAT cases present with the typical features of insidious onset and gradual decline, a high proportion of MID cases lack an abrupt onset and show a more or less gradual decline (Fischer et al., 1989, 1990a). This might be explained by a considerable proportion of silent strokes in MID (Ladurner et al., 1982; Landau, 1989; Chodosh et al., 1988). Daily fluctuations seem to be a rather valid clinical course characteristic in MID (Mölsä et al., 1985; Fischer et al., 1989; Gatterer et al., 1989). The ischemic score of Hachinski is used worldwide to discriminate between MID and DAT clinically, but it heavily overdiagnoses MID (Hachinski et al., 1975; Rosen et al., 1980; Liston and La Rue, 1983a,b; Fischer et al., 1989). The only study which reports a rather high rate of correctly diagnosed MID did not use the ischemic score (Erkinjuntti et al., 1988).

The usefulness of imaging techniques (MRT, SPECT, PET, rCBF) in the differentiation of MID and DAT is uncertain, but SPECT seems to be a reliable and cheap device even in mild stages of dementia in the elderly (Duara et al., 1986; Deutsch and Tweedy, 1987; Erkinjuntti et al., 1987; Jagust et al., 1987; Anderson, 1988; Prohovnik et al., 1988; Johnson et al., 1987, 1988; Burns et al., 1989; Friedland et al., 1989; Leys et al., 1989). Neuropsychology failed to discriminate between MID and DAT (Poeck, 1988; Spinnler and Della Sala, 1988; Marterer et al., 1989; Fischer et al., 1988, 1990b).

The diagnosis of MID and DAT remains a clinical domain and has to take into consideration cognitive, emotional, and behavioural symptoms, and neurological signs, together with an extensive history from a close informant.

References

Alafuzoff I, Iqbal K, Friden H, Adolfsson R, Winblad B (1987) Histopathological criteria for progressive dementia disorders: clinical-pathological correlation and classification by multivariate data analysis. Acta Neuropathol (Berl) 74: 209–225

Alzheimer A (1907) Über eine eigenartige Erkrankung der Hirnrinde. Allgemeine Zeitschrift für Psychiatrie und Psychisch-Gerichtliche Medizin 64: 146–148

Amaducci LA, Rocca WA, Schoenberg BS (1986) Origin of the distinction between Alzheimer's disease and senile dementia. Neurology 36: 1497–1499

American Psychiatric Association (1980) Diagnostic and statistical manual of mental disorders, 3rd edn. Washington DC, APA

American Psychiatric Association (1987) Diagnostic and statistical manual of mental disorders, 3rd edn-revised. Washington DC, APA

Anderson B (1985) MRI and dementia. Neurology 38: 166–167

Berg L (1985) Does Alzheimer's disease represent an exaggeration of normal aging? Arch Neurol 42: 737–739

Berner P (1977) Psychiatrische Systematik, 3. Aufl. Huber, Bern

Bleecker ML, Bolla-Wilson K, Kawas C, Agnew J (1988) Age-specific norms for the Mini-Mental State exam. Neurology 38: 1565–1568

Blessed G, Tomlinson B, Roth M (1968) The association between quantitative measures of dementia and of senile change in the cerebral grey matter of elderly subjects. Br J Psychiatry 114: 797–811

Bleuler E (1916) Lehrbuch der Psychiatrie. Springer, Berlin

Boller F, Lopez OL, Moossy J (1989) Diagnosis of dementia: clinicopathologic correlations. Neurology 38: 76–79

Brink TL, Yesavage JA, Lum O, Heersema PH, Adey M, Rose TL (1982) Screening tests for geriatric depression. Clin Gerontol 1: 37–43

Burns A, Philpot MP, Costa DC, Ell PJ, Levy R (1989) The investigation of Alzheimer's disease with single photon emission tomography. J Neurol Neurosurg Psychiatry 52: 248–253

Caine ED (1981) Pseudodementia. Current concepts and future directions. Arch Gen Psychiatry 38: 1359–1364

Chandra V, Kokmen E, Schoenberg BS, Beard CM (1989) Head trauma with loss of consciousness as a risk factor for Alzheimer's disease. Neurology 39: 1576–1578

Chodosh EH, Foulkes MA, Kase CS, Wolf PA, Mohr JP, Hier DB, Price TR, Furtado JG (1988) Silent stroke in the NINCDS stroke data bank. Neurology 38: 1674–1679

Cooper B, Bickel H (1989) Prävalenz und Inzidenz zu Demenzerkrankungen in der Altenbevölkerung. Nervenarzt 60: 472–482

Crook T, Bartus RT, Ferris SH, Whitehouse P, Cohen GD, Gershon S (1986) Age-associated memory impairment: proposed diagnostic criteria and measures of clinical change — Report of a National Institute of Mental Health Work Group. Dev Neuropsychol 2: 261–276

Cummings JL, Miller B, Hill MA, Neshkes R (1987) Neuropsychiatric aspects of multi-infarct dementia and dementia of the Alzheimer type. Arch Neurol 44: 389–393

Danielczyk W, Fischer P, Gatterer G, Simanyi M (1989) Depression in the course of MID and DAT. J Neural Transm (P-D Sect) 1: 44

Deutsch G, Tweedy JR (1987) Cerebral blood flow in severity-matched Alzheimer and multi-infarct patients. Neurology 37: 431–438

Duara R, Grady C, Haxby J, Sundaram M, Cutler NR, Heston L, Moore A, Schlageter N, Larson S, Rapoport SI (1986) Positron emission tomography in Alzheimer's disease. Neurology 36: 879–887

Erkinjuntti T, Ketonen L, Sulkava R, Sipponen J, Vuorialho M, Iivanainen M (1987) Do white matter changes on MRI and CT differentiate vascular dementia from Alzheimer's disease? J Neurol Neurosurg Psychiatry 50: 37–42

Erkinjuntti T, Haltia M, Palo J, Sulkava R, Paetau A (1988) Accuracy of the clinical diagnose of vascular dementia: a prospective clinical and post-mortem neuropathological study. J Neurol Neurosurg Psychiatry 51: 1037–1044

Ettlin TM, Staehelin HB, Kischka U, Ulrich J, Scollo-Lavizzari G, Wiggli U, Seiler WO (1989) Computed tomography, electroencephalography, and clinical features in the differential diagnosis of sensile dementia. Arch Neurol 46: 1217–1220

Evans DA, Funkenstein HH, Albert MS, Scherr PA, Cook NR, Chown MJ, Hebert LE, Hennekens CH, Taylor JO (1989) Prevalence of Alzheimer's disease in a community population of older persons. JAMA 262: 2551–2556

Farrer LA, O'Sullivan DM, Cupples A, Growdon JH, Myers RH (1989) Assessment of genetic risk for Alzheimer's disease among first-degree relatives. Ann Neurol 25: 485–493

Fillenbaum GG, Heyman A, Wilkinson WE, Haynes CS (1987) Comparison of two screening tests in Alzheimer's disease. The correlation and reliability of the Mini-Mental State examination and the modified blessed test. Arch Neurol 44: 924–927

Fischer P, Gatterer G, Marterer A, Danielczyk W (1988) Nonspecificity of semantic impairment in dementia of Alzheimer's type. Arch Neurol 45: 1341–1343

Fischer P, Jellinger K, Gatterer G, Marterer A, Danielczyk W (1989) Neuropathological validation of the Hachinski Scale. J Neural Transm (P-D Sect) 1: 57

Fischer P, Gatterer G, Marterer A, Simanyi M, Danielczyk W (1990a) Course-characteristics in the differentiation of dementia of Alzheimer's type and multi-infarct dementia. Acta Psychiatr Scand 81: 551–553

Fischer P, Marterer A, Danielczyk W (1990b) Right-left disorientation in dementia of the Alzheimer's type. Neurology 40: 1619–1620

Fischer P, Simanyi M, Danielczyk W (1990c) Depression in dementia of the Alzheimer type and in Multi-Infarct dementia. Am J Psychiatry 147: 1484–1487

Fischer O (1907) Miliare Nekrosen mit drusigen Wucherungen der Neurofibrillen, eine regelmäßige Veränderung der Hirnrinde bei seniler Demenz. Mschr Psychiat Neurol 22: 361–372

Fitch N, Becker R, Heller A (1988) The inheritance of Alzheimer's disease: a new interpretation. Ann Neurol 23: 14–19

Folstein MF, Folstein SE, McHugh PR (1975) "Mini-Mental State". A practical method for grading the cognitive state of patients for the clinician. J Psychiatr Res 12: 189–198

Folstein MF (1983) The Mini-Mental State examination. In: Crook TH, Ferris S, Bartus R (eds) Assessment in geriatric psychopharmacology. Mark Powley, Connecticut, pp 47–51

Friedland RP, Jagust WJ, Huesman RH, Koss E, Knittel B, Mathis CA, Ober BA, Mazoyer BM, Budinger TF (1989) Regional cerebral glucose transport and ultilization in Alzheimer's disease. Neurology 39: 1427–1434

Galasko D, Klauber MR, Hofstetter CR, Salmon DP, Lasker B, Thal LJ (1990) The Mini-Mental State examination in the early diagnosis of Alzheimer's disease. Arch Neurol 47: 49–52

Gatterer G, Fischer P, Danielczyk W (1989) Variation of cognitive parameters in the progress of dementia. J Neural Transm (P-D Sect) 1: 69

Good MI (1981) Pseudodementia and physical findings masking significant. Psychopathology 138: 811–814

Gurland B, Toner J (1987) The epidemiology of the concurrence of depression and dementia. In: Altmann HJ (ed) Alzheimer's disease. Plenum, New York, pp 45–58

Hachinski VC, Iliff LD, Zilhka E, DuBoulay GH, McAllister VL, Marshall J, Russell RWR, Symon L (1975) Cerebral blood flow in dementia. Arch Neurol 32: 632–637

Hachinski VC (1985) Cerebrovascular risk factors. In: Poeck K, Freund H-J, Gänshirt H (eds) Neurology. Springer, Berlin Heidelberg New York, pp 223–228

Hagnell O, Lanke J, Rorsman B, Öjesjö L (1982) Are we entering an age of melancholy? Psychol Med 12; 279–289

Hagnell O, Lanke J, Rorsman B, Öhman R, Öjesjö L (1983) Current trends in the incidence of senile and multi-infarct dementia. Arch Psychiatr Nervenkr 233: 423–438

Hansen L, Salmon D, Galasko D, et al (1990) The Lewy body variant of Alzheimer's disease: a clinical and pathological entity. Neurology 40: 1–8

Henderson AS (1986) The epidemiology of Alzheimer's disease. Br Med Bull 42: 3–10

Henderson AS (1988) The risk factors for Alzheimer's disease: a review and a hypothesis. Acta Psychiatr Scand 78: 257–275

Jaqust WJ, Budinger TF, Reed BR (1987) The diagnosis of dementia with single photon emission computed tomography. Arch Neurol 44: 258–262

Jellinger K (1976) Neuropathological aspects of dementias resulting from abnormal blood and cerebrospinal fluid dynamics. Acta Neurol Belg 76: 83–102

Jellinger K, Danielczyk W, Fischer P, Gabriel E (1990) Clinicopathological analysis of dementia disorders of the elderly. J Neurol Sci 95: 239–258

Joachim CL, Morris JH, Selkoe DJ (1988) Clinically diagnosed Alzheimer's disease: autopsy results in 150 cases. Ann Neurol 24: 50–56

Johnson KA, Mueller ST, Walshe TM, English RJ, Holman BL (1987) Cerebral perfusion imaging in Alzheimer's disease. Arch Neurol 44: 165–168

Johnson KA, Holman BL, Mueller SP, Rosen TJ, English R, Nagel JS, Growdon JH (1988) Single photon emission computed tomography in Alzheimer's disease.

Abnormal Iofetamine I 123 uptake reflects dementia severity. Arch Neurol 45: 392–396

Jorm AF, Korten AE, Henderson AS (1987) The prevalence of dementia: a quantitative integration of the literature. Acta Psychiatr Scand 76: 465–479

Katzman R (1986) Alzheimer's disease. NEJM 314: 964–973

Katzman R (1988) Alzheimer's disease as an age-dependent disorder. In: Research and the ageing population. Wiley, Chichester, pp 69–85 (Ciba Foundation Symposium 134)

Kay DWK (1986) The descriptive epidemiology of the dementias. Neurology 16–33

Kay DWK, Henderson AS, Scott R, Wilson J, Rickwood D, Grayson DA (1985) Dementia and depression among the elderly living in the Hobart community: the effect of the diagnostic criteria on the prevalence rates. Psychol Med 15: 771–788

Kiloh LG (1961) Pseudo-dementia. Acta Psychiatr Scand 37: 336–351

Klvelä SL, Pahkala K, Laippala P (1988) Prevalence of depression in an elderly population in Finland. Acta Psychiatr Scand 78: 401–413

Knesevich JW, Martin RL, Berg L, Danzinger W (1983) Preliminary report of affective symptoms in the early stages of senile dementia of the Alzheimer type. Am J Psychiatry 140: 233–235

Knopman DS, Kitto J, Deinard S, Heiring J (1988) Longitudinal study of death and institutionalization in patients with primary degenerative dementia. JAGS 36: 108–112

Kokmen E, Beard CM, Offord KP, Kurland LT (1989) Prevalence of medically diagnosed dementia in a defined United States population: Rochester, Minnesota, January 1, 1975. Neurology 39: 773–776

Kokmen E (1989) Epidemiology of dementing illness in Rochester, Minnesota, USA. Paper held at the WFN meeting Aging of the Brain and Dementia, Florence, Italy, June 1, 1989

Kral VA (1962) Senescent forgetfulness: benign and malignant. J Can Med Assoc 86: 257–260

Kral VA (1982) Depressive Pseudodemenz und Senile Demenz vom Alzheimer-Typ. Nervenarzt 53: 284–286

Ladurner G, Iliff LD, Lechner H (1982) Clinical factors associated with dementia in ischaemic stroke. J Neurol Neurosurg Psychiatry 45: 97–101

Landau WM (1989) Au clair de lacune: Holy, wholly, holey logic. Neurology 39: 725–730

Lazarus LW, Newton N, Cohler B, Lesser J, Schweon C (1987) Frequency and presentation of depressive symptoms in patients with primary degenerative dementia. Am J Psychiatry 144: 41–45

Leys D, Steinling M, Petit H, Salomez JL, Gaudet Y, Ovelacq E, Vergnes R (1989) Maladie d'Alzheimer: Etude par tomographie d'émission monophotonique (Hm PAO Tc^{99m}). Rev Neurol 145: 443–450

Li G, Shen YC, Che CH, Zhao YW, Li SR, Lu M (1989) An epidemiological survey of age-related dementia in an urban area of Beijing. Acta Psychiatr Scand 79: 557–563

Liston EH, La Rue A (1983a) Clinical differentiation of primary degenerative and multi-infarct dementia: a critical review of the evidence. Part I. Clinical studies. Biol Psychiatry 18: 1451–1465

Liston EH, La Rue A (1983b) Clinical differentiation of primary degenerative and multi-infarct dementia: a critical review of the evidence. Part II. Pathological studies. Biol Psychiatry 18: 1467–1484

Madden JJ, Luhan JA, Kaplan LA (1952) Nondementing psychoses in older persons. JAMA 150: 1567

Marsden CD (1985) Assessment of dementia. In: Frederiks JAM (ed) Handbook of clinical neurology 46: 221–232

Marterer A, Fischer P, Danielczyk W (1989) Ideomotor apraxia in the course of DAT. J Neural Transm (P-D Sect) 1: 100

McAllister TW, Price TRP (1982) Severe depressive pseudodementia with and without dementia. Am J Psychiatry 139: 626–629

McKhann G, Drachman D, Folstein M, Katzman R, Price D, Stadlan EM (1984) Clinical diagnosis of Alzheimer's disease: report of the NINCDS-ADRDA Work Group under the auspices of Department of Health and Human Services Task Force on Alzheimer's disease. Neurology 34: 939–944

Meehl PE, Rosen A (1955) Antecedent probability and the efficiency of psychometric signs, patterns, or cutting scores. Psychol Bull 52: 194–216

Miller DF, Hicks SP, D'Amato CJ, Landis RJ (1984) A descriptive study of neuritic plaques and neurofibrillary tangles in an autopsy population. Am J Epidemiol 120: 331–341

Morris JC, McKeel DW, Fulling K, Torack RM, Berg L (1988) Validation of clinical diagnostic criteria for Alzheimer's disease. Ann Neurol 24: 17–22

Mölsä PK, Marttila RJ, Rinne UK (1982) Epidemiology of dementia in a Finnish population. Acta Neurol Scand 65: 541–552

Mölsä PK, Paljärvi L, Rinne JO, Rinne UK, Säkö E (1985) Validity of clinical diagnosis in dementia: a prospective clinicopathological study. J Neurol Neurosurg Psychiatry 48: 1085–1090

Ng HK, Lee JCK (1988) Degenerative cerebral alterations in Chinese aged 65 years or older. Clin Neuropathol 7: 280–284

Pearlson GD, Rabins PV, Kim WS, Speedie LJ, Moberg PJ, Burns A, Bascom MJ (1989) Structural brain CT changes and cognitive deficits in elderly depressives with and without reversible dementia ('pseudodementia'). Psychol Med 19: 573–584

Perry RH, Irving D, Blessed G, Fairbairn A, Perry EK (1990) Senile dementia of Lewy body type. J Neurol Sci 95: 119–139

Pettinati HM, Brown M, Mathisen KN (1985) Memory complaints in depressed geriatric inpatients. Ann NY Acad Sci 444: 528–530

Poeck K (1988) A case for neuropsychology in dementia research. J Neurol 235: 257

Prohovnik I, Mayeux R, Sackeim HA, Smith G, Stern Y, Alderson PO (1988) Cerebral perfusion as a diagnostic marker of early Alzheimer's disease. Neurology 38: 931–937

Rabins PV, Merchant A, Nestadt G (1984) Criteria for diagnosing reversible dementia caused by depression: validation by 2-year follow- up. Br J Psychiatry 144: 488–492

Reifler BV, Larson E, Hanley R (1982) Coexistence of cognitive impairment and depression in geriatric outpatients. Am J Psychiatry 139: 623–626

Reisberg B, Borenstein J, Salob SP, Ferris SH, Franssen E, Georgotas A (1987) Behavioral symptoms in Alzheimer's disease: phenomenology and treatment. J Clin Psychiatry 48: 9–15

Rocca WA, Amaducci LA, Schoenberg BS (1986) Epidemiology of clinically diagnosed Alzheimer's disease. Ann Neurol 19: 415–424

Rosen WG, Terry RD, Fuld PA, et al (1980) Pathological verification of ischemie score in differentiation of dementias. Ann Neurol 7: 486–488

Roth M (1955) The natural history of mental disorder in old age. J Ment Sci 101: 281–301

Schoenberg BS (1986) Descriptive epidemiology of Parkinson's disease: disease distribution and hypothesis formulation. Adv Neurol 45: 277–283

Sheikh JI, Yesavage JA (1986) Geriatric depression scale (GDS). Recent evidence and development of a shorter version. Clin Gerontol 5: 165–173

Shibayama H, Kasahara Y, Kobayashi H, et al (1986) Prevalence of dementia in a Japanese elderly population. Acta Psychiatr Scand 74: 144–151

Siegel B, Gershon S (1987) Dementia, depression, and pseudodementia. In: Altman HJ (ed) Alzheimer's disease problems, prospects, and perspectives. Plenum Press, New York, pp 29–44

Simchowicz T (1911) Histologische Studien über die senile Demenz. In: Nissl F, Alzheimer A (Hrsg) Histologische und histopathologische Arbeiten über die

Großhirnrinde mit besonderer Berücksichtigung der pathologischen Anatomie der Geisteskranken. Gustav Fischer, Jena, S267–443

Spinnler H, Della Sala S (1988) The role of clinical neuropsychology in the neurological diagnosis of Alzheimer's disease. J Neurol 235: 258–271

Streifler M, Simanyi M, Fischer P, Danielczyk W (1990) EEG- and cognitive changes in Alzheimer's disease (AD) and senile dementia of AD-type (SDAT). In: Beckman H, Maurer K, Riederer P (eds) Alzheimer's disease: epidemiology, neuropathology, neurochemistry and clinics. Springer, Wien New York, pp 447–458 (Key Topics in Brain Research)

Strömgren LS (1977) The influence of depression on memory. Acta Psychiatr Scand 56: 109–128

Sulkava R, Haltia M, Paetau A, Wikström J, Palo J (1983) Accuracy of clinical diagnosis in primary degenerative dementia: correlation with neuropathological findings. J Neurol Neurosurg Psychiatry 46: 9–13

Sunderland T, Alterman IS, Yount D, Hill JL, Tariot PN, Newhouse PA, Mueller EA, Mellow AM, Cohen RM (1988) A new scale for the assessment of depressed mood in demented patients. Am J Psychiatry 145: 955–959

Thal LJ, Grundman M, Golden R (1986) Alzheimer's disease: a correlational analysis of the Blessed information-memory-concentration test and the Mini-Mental State examination. Neurology 36: 262–264

Tierney MC, Fischer RH, Lewis AJ, Zorzitto ML, Snow WG, Reid DW, Nieuwstraten P (1988) The NINCDS-ADRDA Work Group criteria for the clinical diagnosis of probable Alzheimer's disease: a clinicopathologic study of 57 cases. Neurology 38: 359–364

Tomlinson BE, Blessed G, Roth M (1970) Observations on the brains of demented old people. J Neurol Sci 11: 205–242

Tuszynski MH, Petito CK, Levy DE (1989) Risk factors and clinical manifestations of pathologically verified lacunar infarctions. Stroke 20: 990–999

Venna N, Mogocsi S, Jay M, Phull B, Ahmed I (1988) Reversible depression in Binswanger's disease. J Clin Psychiatry 49: 23–26

Weingartner H, Cohen RM, Murphy DL, Martello J, Gerdt C (1981) Cognitive processes in depression. Arch Gen Psychiatry 38: 42–47

Weiss IK, Nagel CL, Aronson MK (1986) Applicability of depression scales to the old person. J Am Geriatr Soc 34: 215–218

Wells CE (1979) Pseudodementia. Am J Psychiatry 136: 895–900

Wragg RE, Jeste DV (1989) Overview of depressive and psychosis in Alzheimer's disease. Am J Psychiatry 146: 577–587

Yesavage JA, Brink TL (1983) Development and validation of geriatric depression screening scale: a preliminary report. J Psychiatry Res 17: 37–49

Zubenko GS, Moossy J (1988) Major depression in primary dementia. Clinical and neuropathologic correlates. Arch Neurol 45: 1182–1186

Authors' address: Dr. P. Fischer, Neurologisches Institut, Schwarzspanierstraße 17, A-1090 Wien, Austria

J Neural Transm (1991) [Suppl] 33: 49–52
© by Springer-Verlag 1991

Multiinfarct dementia

H. Lechner and **G. Bertha**

Department of Neurology and Psychiatry, University of Graz, Austria

Summary. Clinicians have long recognized that dementia is a common symptom among the elderly. The diagnosis of dementia requires us to document the individual's current level of mental functioning and some higher level of intellectual function in the past. The recognition of early or mild cases is specially difficult. Out of epidemiological studies it has been shown that the incidence for multi-infarct dementia (MID) increases with age and is slightly higher among men.

Introduction

Studies from Europe, North America and Asia indicate that multiinfarct dementia is a leading cause of dementing illness among Oriental poulations and a second leading cause among Caucasians. According to the U.S. Bureau of the Census, the age group of 65 and over is expected to increase in size moving from 25 million in 1980 to an estimated 32 million in year 2000 — similar trends are expected for other developed countries (Schoenberg, 1988). Thereby the part of population at higher risk for vascular dementia is expected to increase.

Clinical picture

The onset of vascular dementia is frequently sudden or acute and dementia often will become apparent after the occurrence of a stroke. MID patients frequently express distress about their loss of memory before the first stroke is turning up. Vascular dementia may also be found without any clinical dates of previous cerebrovascular events. Several studies have concluded that impairment of intellectual functions despite complete recovery of other neurological signs are a frequent occurrence in patients suffering from transient ischemic attacks (Wolf et al., 1988; Loeb, 1988). The clinical symptoms of dementia will sometimes be concealed by a depressive state associated with anxiety, restlessness and lack of interest and will be worsened by additional psychosocial stress. Features like loss or defect in coping with the events of daily living, difficulties in using public transportation, managing own financial affairs and perform official acts are often the first signs of psychic decompensation. The course of the disease is characterised not by changing the symptomatology but by loss of combined functions of social integration. So changing patterns of working capability among our patients throughout the observationperiod of five years could be observed. Thus, nearly all had to retire due to the disease. People with a higher educational level and self employed people remain longer at work (Lechner et al., 1988). The disease is therefor of big social interest, as someone has to carry the burden of care. There will be contrasts in different countries depending from social wellfare state, people's attitude to family life and also underlying religion may play an important role.

The prognosis of MID is determined by the fact that out of 88 patients 38 (43%) died during the follow up period and the cause of death was in two thirds of vascular origin. Further it could be shown that the later the MID starts in late life the worse the prognosis is. Also nocturnal confusion is a bad prognostic sign for outlive (Lechner et al., 1988).

For the diagnosis of vascular dementia the Hachinski ischemic score (Hachinski et al., 1975) has been confirmed by autopsy as most valuable. A high ischemic score identified those dementias with vascular disease (Molsa et al., 1985) and distinguished them from cases of senile dementia of Alzheimer's type (SDAT). But up till now the problem arises as dementia of Alzheimer's type does not exclude to have an additional cerebraovascular disease.

The different neuropsychological profile seems to be not to helpful to distinguish cortical from subcortical dementia. Cortical dementia is said to consist of aphasia, apraxia and visuaspatial problems in contrast to the mental slowness of the subcortical dementia associated with marked motor problems (Wells, 1980; Harrison, 1988).

The differential diagnosis had been greatly advanced by single photon emission computed tomography (SPECT) and magnetic resonance imaging (MRI). Also brain mapping can be useful in the differential diagnosis of dementia, whereby patients with vascular dementia showed significantly higher amplitudes in the delta and theta bands, the alpha frequencies and amplitudes were significantly lower in the SDAT group. Duplex and transcranial Doppler sonography is not to helpful as they are showing only the

Table 1. Comparison of SPECT results in patients with SDAT and vascular dementia (VD)

Cortical minorperfusion	SDTA (N = 19)	VD (N = 17)
Normal	0	6 (35%)**
Frontal symmetrically	5 (26%)	2 (12%)
Parietotemporal symmetrically	14 (73%)	1 (6%)***
Patchy	1 (5%)	8 (47%)**

Fisher's exact test: ** p < 0.01; *** p < 0.001

Table 2. MRI findings in normal aging, SDAT and VD

	Normals (N = 32) N (%)	SDAT (N = 17) N (%)	VD (N = 19) N (%)
Cerebral infarcts	1 (3)	0 (0)	14 (73)** ††
BGH	9 (28)	4 (23)	16 (84)** †††
BGL	0 (0)	0 (0)	9 (47)** ††
Confluent WMH	1 (3)	0 (0)	8 (42)*** ††
Uncal-hippocampal and/or insular ICSI	7 (21)	10 (59)* ††	2 (11)
Cortical atrophy	7 (21)	17 (100)***	15 (88)***
Ventricular atrophy	3 (9)	16 (94)***	15 (88)***

* vs. controls † SDAT vs. VD (1 symbol p < 0.05; 2 symbols p < 0.01; 3 symbols p < 0.001)
BGH basal ganglia hyperintensities
BGL basal ganglia lacunes
WMH white matter hyperintensities
ICSI increased cortical signal intensity
(Schmidt et al., 1989)

involvement of extra- and intracranial vessels. Difference could be seen in SPECT (Lechner et al., 1990; Table 1).

Most helpful information can be got out of the MRI. The differential diagnosis can be passed on the incidence of cerebral infarcts, basal ganglia lacunes, basal ganglia hyperintensities, confluent white matter hyperintensities and uncal, hippocampal and/or insular increased cortical signal intensity (Table 2).

References

Hachinski VC, Iliff LD, Zilkha E, et al (1975) Cerebral blood flow in dementia. Arch Neurol 32: 632–637

Harrison MJG (1988) Clinical features of vascular and multi-infarct dementia. In: Meyer JS, Lechner H, Marshall J, Toole JF (eds) Vascular and multi-infarct dementia. Futura Publishing Co, Mount Kisco NY, pp 5–12

Lechner H, Bertha G, Ott E (1988) Results of a five-year prospective study of 94 patients with vascular and multi-infarct dementia. In: Meyer JS, Lechner H, Marshall J, Toole JF (eds) Vascular and multi-infarct dementia. Futura Publishing Co, Mount Kisco NY, pp 101–111

Lechner H, Niederkorn K, Logar C, Schmidt R, Fazekas F (1990) EEG-brain mapping in patients with SDAT and vascular dementia — results and correlations with MRI, SPECT and TCD. In: Battistin L (ed) Aging brain and dementia: new trends in diagnosis and therapy. Alan R Liss, New York, pp 337–348

Loeb C (1988) Intellectual function, transient ischemic attacks, and vascular and multi-infarct dementia. In: Meyer JS, Lechner H, Marshall J, Toole JF (eds) Vascular and multi-infarct dementia. Futura Publishing Co, Mount Kisco NY, pp 23–33

Mirsen T, Hachinski V (1988) Epidemiology and classification of vascular and multi-infarct dementia. In: Meyer JS, Lechner H, Marshall J, Toole JF (eds) Vascular and multi-infarct dementia. Futura Publishing Co, Mount Kisco NY, pp 61–76

Molsa PK, Paljarvi L, Rinne JO, et al (1985) Validity of clinical diagnosis in dementia: a prospective clinicopathological study. J Neurol Neurosurg Psychiatry 48: 1085–1089

Schmidt R, Fazekas F, Offenbacher H, et al (1989) A comparison of MRI in normal aging, senile dementia of Alzheimer's type (SDAT) and vascular dementia (VD). Book of abstracts, vol. 1. Society of Magnetic Resonance in Medicine. 8th Annual Meeting and Exhibition, August 12–18, 1989, Amsterdam

Schoenberg BS (1988) Epidemiology of vascular and multi-infarct dementia. In: Meyer JS, Lechner H, Marshall J, Toole JF (eds) Vascular and multi-infarct dementia. Futura Publishing Co, Mount Kisco NY, pp 47–59

Wells CE (1980) The differential diagnosis of psychiatric disorders in the elderly. In: Cole JO, Barrett JE (eds) Psychopathology in the aged. Raven Press, New York, pp 19–31

Wolf PA, Kase CS, Cupples LA, Bachman DL, Dyken ML, Yatsu FM (1988) Epidemiology of TIAs and the possibilities of preventing vascular and multi-infarct dementia. Part I–IV. In: Meyer JS, Lechner H, Marshall J, Toole JF (eds) Vascular and multi-infarct dementia. Futura Publishing Co, Mount Kisco NY, pp 77–99

Authors' address: Prof. Dr. Dr. hc. H. Lechner, Department of Neurology, University of Graz, Auenbruggerplatz 22, A-8036 Graz, Austria

J Neural Transm (1991) [Suppl] 33: 53–58
© by Springer-Verlag 1991

Early diagnosis of Alzheimer dementia?

C. Lind, M. Mraz, C. Wöber, I. Marschall, L. Deecke, and **P. Dal-Bianco**

Neurology University Clinic, Vienna, Austria

Summary. The main problems in early diagnosis of Alzheimer dementia (AD) are:

1. The differentiation between normal aging and AD i.e. difficulties in the assessment of cognitive disturbances in the healthy elderly and in early demented subjects.

2. Interference with other dementia syndromes.

3. Lack of information in the population and among physicians about the different causes and courses of dementia syndromes. The first two aspects are discussed in this paper.

Introduction

The following article deals with the difficulties in early diagnosis of AD focussing on symptomatology and neuropsychological testing. In the first part the relationship normal aging — AD is illuminated. The second part is about age-associated-memory-impairment and its relation to early dementia. It presents also longitudinal studies on the course and the development of dementia. The last part refers to pseudodementia and the difficulties in distinguishing it from AD.

Normal aging and dementia

The differentiation between normal brain aging and early dementia remains controversial. Pathohistological changes are qualitatively similar in both processes. However there are topographical and quantitative differences, leading to the assumption that there must be an association between them, perhaps in the way that different etiological factors lead to a common final pathway of degeneration (Landfield, 1986). AD would therefore represent a kind of "activation and acceleration of the normal machinery of brain aging" by genetic, exogenous and other factors. However the interaction of these factors is very difficult to assess and has to be clarified in further investigations (Creasey and Rapoport, 1985; Rabitt, 1983).

The relationship between normal aging and dementia may be causal, contributory or coincidental (Drachman, 1983): the causal hypothesis establishes a direct connection between normal aging and dementia. The contributory theory assumes that dementia is only partly due to normal aging and that the deficits of normal aging are increased by additional factors (such as cerebrovascular or traumatic events) or accelerated by genetic and enviromental conditions. Genetic factors, for example, play an important role in early onset AD. The coincidental hypothesis suggests that specific disease processes unrelated to normal aging, e.g. slow virus infection, lead to the development of dementia, this theory being less convincing than the two cited above. Another author points out some important differences between normal aging and AD (Berg, 1985): first, there is no increased amount of lipofuscin in neurons of AD brains in contrary to the findings in "normal" old brains. Secondly, loss of dendritic arborisation is found in AD brains, but not in normal aging.

Seen from the epidemiological point of view, it is possible to assume that normal aging and AD lie on a continuum when observing the frequency distribution of some of the variables associated with AD e.g. cognitive test scores and behaviour scores. They appear to be unimodal, continuous and highly skewed (Brayne and Calloway, 1988) leading the authors to the provocative statement that one should move away from the model of "has he got it?" to that of "how much of it has he got and why?"

Age-associated memory impairment and dementia

In order to evaluate cognitive changes and especially memory impairment in healthy old people the term age-associated-memory-impairment (AAMI) was introduced some years ago (Crook and Bartus, 1986; Crook and Larrabee, 1988). Those individuals who perform at one SD below the average for the young population and who complain that their memory has declined are defined as meeting the criteria for AAMI (see Fig. 1). Applying the inclusion and exclusion criteria for this complex, almost half of the normal elderly population seems to fall in this category (Ferris and Flicker, 1989). Those subjects formerly described having benign senescent forgetfulness (BSF) (Kral, 1962) obviously belong to the more severely impaired end of the bell shaped AAMI distribution and represent borderline dementia (see Fig. 1). The term BSF has been questioned since a prospective longitudinal study (Katzman and Aronson, 1989) showed that 37% of these subjects developed dementia.

When comparing the cognitive decline in normal aging and early dementia the differences are mainly quantitative. It is very important, however, to emphasize the fact that there are some factors, e.g. immediate memory that are unimpaired in both groups whereas naming and remote memory impairments appear in early AD but not in normal aging (Ferris and Flicker, 1989). A special test battery has been developed testing the ability to perform everyday relevant tasks and to eliminate investigator related interference

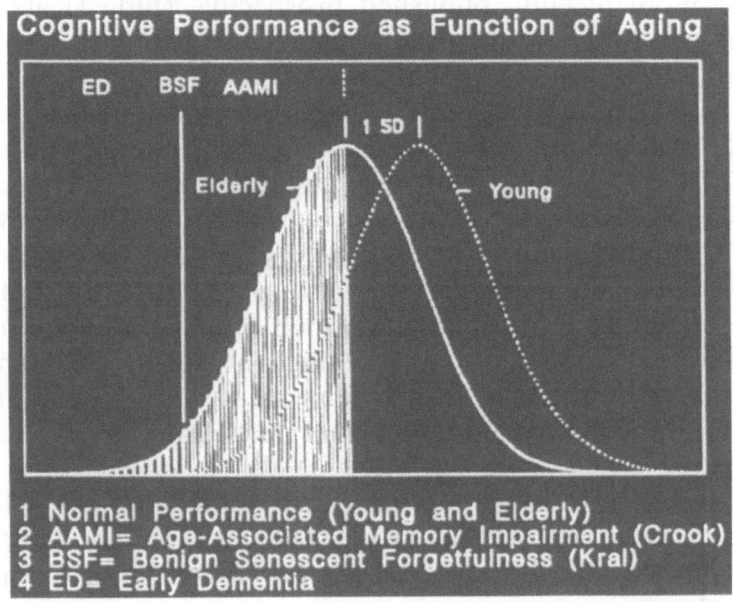

Fig. 1. Individuals who perform at one SD below the average for the young population and who complain that their memory has declined are defined as meeting the criteria for AAMI. Subjects with benign senescent forgetfulness or early dementia belong to the severely impaired end of the bell shaped AAMI distribution (modified after Crook, 1986)

factors (Crook and Bartus, 1986; Crook and Larrabee, 1988; Ferris and Flicker, 1989).

A practical method for the clinical classification of dementia patients is the clinical dementia rating scale (CDR) (Berg et al., 1982, 1984, 1988). It comprises 6 items (memory, orientation, judgement + problem solving, community affairs, home + hobbies, personal care) and has proved valuable for classifying the degree of dementia (CDR1–CDR3), CDR 0,5 representing the borderlands of questionable dementia which may include patients with early AD or patients with BSF. What clearly distinguishes healthy elderly from AD patients is that the former can be forgetful in everyday activities — e.g. names and dates, but their forgetfulness is not constant, does not interfere with everyday life and is of no major concern to them (Berg et al., 1982).

An important question is whether it is possible to predict the degree of dementia after a certain time interval, based on the data collected at the beginning of the observation. Berg et al. (1984) came to the conclusion that the main indicators of the severity of dementia after a certain time (in this case 1 year) turned out to be the estimated duration of illness, the brief clinical scales, the six rating categories of the CDR and most of the psychometric measures whereas EEG, VEP or CT were not helpful in predicting the course of the illness.

Another more recently published prospective study (Katzman et al., 1989) was done to observe the development of dementia in a group of non-demented old people aged 75 to 85.

56 out of 434 volunteers developed progressive dementia, 32 of them meeting diagnostic criteria for AD. The main risk factors for developing AD found in this group were age (>80) and gender (female). The major predictor of dementia was the mental status score in the Blessed IMC test on entry. Most of the subjects with more than five errors on the IMC-test at the beginning of the study made errors in all or parts of the five point memory phrase question and 37% of them developed dementia as mentioned above. Consequently they should not be classified as having "benign senescent forgetfulness". It was also shown in this study that depression does not usually occur at the early phase of AD but it is found in MID and MID/AD patients.

There have been trials to find out which neuropsychological test would give good diagnostic accuracy in early AD stages (Spinnler and Della Salla, 1988; Storandt et al., 1984; Tierney et al., 1987). One group (Storandt et al., 1984) filtered out four tests of a big neuro-psychological test battery using discriminant function analysis. The diagnostic accuracy reached 98% with the combination of the logical memory and mental control subtest of the Wechsler Memory Scale, Form A of the Trailmaking test and word fluency for letters S and P. These results were confirmed by another group (Tierney et al., 1987) who added the finding that these tests are not so valuable in differentiating different types of dementia.

What are the outstanding features of cognitive changes in early AD that can be reported by relatives and persons close to AD patients? There is a well documented case of an 82 year old healthy man who was assessed three times and at the third evaluation was classified as being questionably demented (CDR 0,5) (Morris and Fulling, 1988). At that time his wife described a subtle intellectual loss in the preceeding 6 months and characterized his thinking and memory as "vague". There were difficulties in remembering appointments, he attended less well to financial details and needed some help in orientation. At that time he denied problems with memory or thinking. He made errors in neuropsychological testing, i.e. in recent memory and in the face-hand-test. After death autopsy confirmed AD.

Pseudodementia

One of the most confusing syndromes that interferes with the diagnosis of AD is that of pseudodementia (Caine, 1981; Guterman and Eisdorfer, 1989; Knesevich et al., 1983).

It may be almost impossible to differentiate dementia from depressive pseudodementia on clinical grounds even by neuropsychological testing. Additionally there are cases where both disease processes appear in the same subject. An indication towards depression as an underlying cause is the inconsistent test performance of individuals who are tested several times.

They are mostly impaired in attention, mental processing speed, spontaneous elaboration and analysis of detail (Caine, 1981). The impairment is probably due to deficits in attention-arousal-concentration, spontaneous verbal elaboration and rapid, detailed analysis functions which are not primarily cortically mediated.

It an earlier conducted study (Kahn et al., 1975) it has been shown that the complaint about memory function is related to the level of depression but does not correspond to the actual test performance, whereas complaints are minimal in subjects with organic brain disease.

Early diagnosis of AD?

Considering some of the aspects cited above, it seems to be possible to detect those subjects who are at risk of developing dementia and who have to be examinated carefully and observed longitudinally in order to confirm the initial diagnosis of early AD or to relate the process to other disorders especially when aiming at therapeutic consequences which may only be useful at the beginning of the disease.

References

Berg L, Hughes CP, et al (1982) Mild senile dementia of Alzheimer type: research diagnostic criteria, recruitment and description of a study population. J Neurol Neurosurg Psychiatry 45: 962–968

Berg L, Miller JP, et al (1988) Mild senile dementia of the Alzheimer type. 2. Longitudinal assessment. Ann Neurol 23: 477–484

Berg L, Danziger WL, et al (1984) Predictive features in mild senile dementia of the Alzheimer type. Neurology 34: 563–569

Berg L (1985) Does Alzheimer's disease represent an exaggeration of normal aging? Arch Neurol 42: 737–739

Brayne C, Calloway P (1988) Normal ageing, impaired cognitive function, and senile dementia of the Alzheimer's type: a continuum? Lancet: 1265–1267

Creasey H, Rapoport SI (1985) The aging human brain. Ann Neurol 17: 2–10

Caine ED (1981) Pseudodementia — current concepts and future directions. Arch Gen Psychiatry 38: 1359–1364

Crook Th, Bartus RT, et al (1986) Age-associated memory impairment: proposed diagnostic criteria and measures of clinical change-Report of a National Institute of Mental Health Work Group. Dev Neuropsychol 2(4): 261–276

Crook TH, Larrabee GJ (1988) Interrelationships among everyday memory tests: stability of factor structure with age. Neuropsychology 2: 1–12

Drachman DA (1983) How normal aging relates to dementia: a critique and classification. In: Samuel D, et al (eds) Aging of the brain. Raven Press, New York, pp 19–31

Ferris SH, Flicker C, et al (1989) Age-associated memory impairment, benign forgetfulness dementia. In: Bergener, Reisberg (eds) Diagnosis and treatment of senile dementia. Springer, Berlin Heidelberg New York Tokyo, pp 72–82

Guterman A, Eisdorfer C (1989) Early diagnosis of dementia. In: Bergener, Reisberg (eds) Diagnosis and treatment of senile dementia. Springer, Berlin Heidelberg New York Tokyo, pp 177–192

Kahn RL, Zarit SH, et al (1975) Memory complaint and impairment in the aged. Arch Gen Psychiatry 32: 1569–1573

Katzman R, Aronson M, et al (1989) Development of dementing illness in an 80-year-old volunteer cohort. Ann Neurol 25: 317–324

Knesevich JW, Martin RL, et al (1983) Preliminary report on affective symptoms in the early stages of senile dementia of the Alzheimer type. Am J Psychiatry 140: 233–235

Kral VA (1962) Senesccent forgetfulness: benign and malignant. Can Med Assoc J 86: 257–260

Landfield Ph W (1986) Preventive approaches to normal brain aging and Alzheimer's disease. In: Crook, Bartus, et al (eds) Treatment development strategies for Alzheimer's disease. Mark Powley, Madison Connecticut, pp 221–244

Morris JC, Fulling K (1988) Early Alzheimer's disease-diagnostic considerations. Arch Neurol 45: 345–349

Rabbitt P (1983) How can we tell whether human performance is related to chronological age? In: Samuel D, et al (eds) Aging of the brain. Raven Press, New York, pp 9–19

Spinnler H, Della Salla S (1988) The role of clinical neuropsychology in the neurological diagnosis of Alzheimer's disease. J Neurol 235: 258–271

Storandt M, Botwinick J, et al (1984) Psychometric differentiation of mild senile dementia of the Alzheimer type. Arch Neurol 41: 497–499

Tierney MC, Snow G, et al (1987) Psychometric differentiation of dementia-replication and extension of the findings of Storandt and coworkers. Arch Neurol 44: 720–722

Authors' address: C. Lind, MD, Neurologische Universitätsklinik, Lazarettgasse 14, A-1097 Wien, Austria

J Neural Transm (1991) [Suppl] 33: 59–63
© by Springer-Verlag 1991

Galanthamine treatment in Alzheimer's disease

P. Dal-Bianco, J. Maly, Ch. Wöber, C. Lind, G. Koch, J. Hufgard, I. Marschall, M. Mraz, and **L. Deecke**

Neurology University Clinic, Vienna, Austria

Summary. 18 patients who had fullfilled the NINCDS-ADRDA criteria for "possible AD" took part in a clinical study to evaluate the effect of the cholinesteraseinhibitor Galanthamine, 30 mg/day. Neuropsychological und social parameters were rated. This open clinical pilot-study showed no statistic significant change in neuropsychological test-results. However after 1 year treatment 6 patients are still taking the drug. According to their care-persons there was a positive changes in competence of everyday-routine and/or in the emotional situation.

In general, the neuropathological changes in patients suffering form Alzheimer's disease (AD) are an exaggeration of those found during normal ageing: Plaques, tangles and cerebral atrophy occur in both cases (McKhann et al., 1984).

A major difference appears to be the loss of cholinergic neurons in the nucleus basalis of Meynert in patients with AD (Whitehouse et al., 1981). The degeneration of these cholinergic neurons, which innervate the cerebral cortex and hippocampus, is closely linked with the decrease of cholineacetyl-transferase (CAT) (Coyle et al., 1983; Davies et al., 1976).

The decrease in CAT-activity has been correlated with the decline in mental function in senile dementia (Perry et al., 1978).

These findings suggest that enhancement of cholinergic synaptic transmission in the cerebral cortex and hippocampus may provide treatment for patients with AD (Davis et al., 1986). In 1986 Summers et al. published the result of a 3 phase study of oral THA (Tacrine, Tetrahydroaminoacridine) treatment in 17 AD patients. Inspite of promising study-results, changes in liver chemistry put the clinical trial of THA temporary on hold. In 1987 we started an open THA-study. 6 AD patients were treated over a period of 3 to 24 weeks. Patients with mild AD showed some improvement in the neuro-psychological score, whereas moderate AD patients did not improve. However we had to stop our study, because patients treated with a dosage of 100 mg THA or more orally per day showed an elevation of liver enzyme transaminases. The initially promising results with THA resulted in the search for another, but nontoxic centrally active anticholinesterase.

Table 1. Neuropsychological testbattery

Verbal performance
General memory score
Associative memory score
Word list score
Number list score
Remember numbers score

Non-verbal performance
Picure-recognition score
Visual memory score

Language
Aphasia score
Designation score
Fluidity of words score
Vocabulary score

Motor performance
Mosaic test
Reaction time
Tapping frequency
Labyrinth time
Labyrinth touches
Labyrinth error time
Tactile test

Attention
Flicker frequency
Assessing a minute

Galanthamine is a tertiary amine of the phenanthrene group. It is a long acting anticholinesterase and has been used for many years in post-anesthesia without signs of toxicity (Westra et al., 1986). After administration, the cerebral concentration is more than 3 times higher than the plasma peak level. The elimination half-time is 6.8 hrs. (Thomsen et al., 1991) and therefore much longer than that of THA or physostigmine. Positive effects of Galanthamine on spatial memory in mice were reported (Sweeney et al., 1988).

The aim of the open clinical pilot study was to evaluate the effect of Gelanthamine (30 mg/day) on AD patients. Following neuropsychological (Table 1) and social parameters were rated:

— verbal, nonverbal, language, motor and attention performances;
— competence of every-day-routine, emotional situtation, eating habits, toiletting and dressing ability.

18 patients who had fullfilled the NINCDS-ADRDA criteria for "possible AD" took part in the study. Their ages were between 53 and 83 years. They scored at stage I to IV according to Summers et al. (1981).

After 2 months treatment 10 AD patient continued the Galanthamine-study for another 4 months. 6 of these patients have been taking the drug now for 13 to 16 months.

This open clinical pilot-study showed no statistic significant change in neuropsychological test-results in the AD patients treated with Galanthamine for 2 and 6 months.

However after 1 year treatment 6 patients are still taking the drug (Table 2). According to their care-persons there was a positive changes in competence of everyday-routine and/or in the emotional situation while taking the drug and worsening during drug-break (Table 3).

No toxicity or major clinical side-effects were noticed during treatment with Galanthamine in dosages between 30 and 50 mg per day.

Conclusion

— It would appear that those patients responding to Galanthamine treatment belong to a clinical subgroup of Alzheimer disease
— The fact that drug holiday worsens the patient's condition speaks for drug efficacy. According to the care persons even a 3 day-drug holiday worsened the care conditions.
— The required daily dosage would appear to be more than 30 mg/day. Patients taking higher dosages (40 or 50 mg) showed the best results in this group according to pharmacological results of Thomsen et al. (1991).
— More emphasis when rating drug efficacy should be laid on the ability of the patient to cope with the everyday-routine and his emotional situation, rather than his performance in abstract neuropsychological tests.

Table 2. Change after 1 year galanthamine treatment

	S.H.	N.I.	S.L.	S.A.	B.I.	W.K.
Age, Sex	54 f	59 f	74 m	68 f	60 f	68 m
AD-duration a	2	6	4	3	5	6
AD-stage (Summers)	I	IV	III	IV	IV	IV
Dose in mg/day	50	40	30	30	30	30
Competence of everyday-routine	+	+	+	−	−	−
Emotional situation	+	+	+	+	−	+
Eating,	+	+	−	−	+	+
toiletting,	+	+	−	+	−	+
dressing	+	+	+	+	+	−
N.PSY.-test performance	+	−	+	−	−	+

+ same or better
− worse

Table 3. Case reports Pat./daily dosage

	Improvement	Worsening
S.H. 50 mg	householdactivity travels with husband	"hotel-orientation" drugbreak → worsening
N.I. 40 mg	householdactivity orientation in home street building emotional stability	3 times drugbreak → worsening
S.L. 30 mg	orientation in building emotional stability	
S.A. 30 mg	orientation on street	emotional instable worsening after drug-break
B.I. 30 mg	emotional stability	orientation
W.K. 30 mg	orientation in flat emotional stability	dressing

However, to determine to what extent improvement was due to management of patients, to treatment with Galanthamine or both, a double-blind, placebo-controlled study of drug responders would have to be made.

References

Coyle JT, Price DL, DeLong MR (1983) Alzheimer's disease: a disorder of cortical cholingergic innervation. Science 219: 1184–1190

Dal-Bianco P, Maly J, Deecke L (1988) THA-Therapie bei Patienten mit seniler Demenz vom Alzheimer-Typ. Wien Klin Wochenschr 100(12): 415

Davies P, Maloney AJF (1976) Selective loss of central cholinergic neurons in Alzheimer's disease. Lancet ii: 1403

Davis KL, Mohs RC (1986) Cholinergic drugs in Alzheimer's disease. N Engl J Med 315: 1286–1287

McKhann G, Drachman D, Folstein M, et al (1984) Clinical diagnosis of Alzheimer's disease. Neurology 34: 939–944

Perry EK, Tomlinson BE, Blessed G, et al (1978) Correlation of cholinergic abnormalities with senile plaques and mental test scores in senile dementia. Br Med J 2: 1457–1459

Summers WK, et al (1981) Use of THA in treatment of Alzheimer-like dementia: pilot study in twelve patients. Biol Psychiatry 16(2): 145

Summers WK, Majovski LV, Marsh GM, et al (1986) Oral THA in long-term treatment of senile dementia, Alzheimer type. N Engl J Med 315: 1241–1245

Sweeney JE, Höhmann CF, Moran TH, Coyle JT (1988) A long-acting cholinesterase inhibitor reverses spatial memory deficits in mice. Pharmacol Biochem Behav 31: 141–147

Thomsen T, Bickel U, Fischer JP, Kewitz H (1991) Tolerance and inhibition of cholinesterases by Galanthamine in healthy volunteers and Alzheimer patients (in press)

Westra P, et al (1986) Pharmacokinetics of galanthamine (a long-acting anticholinesterase drug) in anaesthetized patients. Br J Anaesth 58: 1303–1307

Whitehouse PJ, Price Dl, Clark AW, et al (1981) Alzheimer disease: evidence for selective loss of cholinergic neurons in the nucleus basalis. Ann Neurol 10: 122–126

Authors' address: Dr. P. Dal-Bianco, Neurological University Clinic, Lazarettgasse 14, A-1090 Vienna, Austria

J Neural Transm (1991) [Suppl] 33: 65–71
© by Springer-Verlag 1991

Human blood platelet as research tool in neuropsychopharmacology

P. Bongioanni, F. Dadone, and **M. Donato**

Scuola Superiore di Studi Universitari e di Perfezionamento "S. Anna", Pisa, Italy

Summary. The use of blood platelets as a nerve terminal model for serotonin is well documented. However, it is clear that the use of platelets as a model can be justified only for those parameters where it may be shown that blood platelets and neural cells share almost identical features. The excellent similarity between the serotonin transport mechanisms in platelets and in nerve terminals, and the existence of various receptors for biogenic amines, peptides and substances with neuronal activity on platelet membrane offer a really unique opportunity to utilize blood platelets as a system for drug evaluation. In our work platelet benzodiazepine binding sites and their modulation by different benzodiazepines in normals and in demented patients are examined.

Introduction

Neurochemical and neuropharmacological studies carried out extensively in a wide range of CNS preparations from experimental animals have contributed greatly to the elucidation of the mechanisms of chemical neurotransmission in human brain, both in healthy and in pathological states. However, extrapolating results to the human condition requires that the processes under study are equivalent in the two species, or may be normalized by reference to a control parameter. An alternative to studies of animals models of human diseases relies on the use of easily obtained human peripheral tissues, such as the formed elements in the blood, which may serve as model systems for studying CNS biochemical processes. Enzyme activities and uptake mechanisms have been studied in human blood platelets, where such functions appear to be more similar to their CNS counterparts than are those of leukocytes or erythrocytes: this may reflect the common embryological origin of platelets and neurones, a contention strengthened by the finding that neurone-specific enolase (NSE) is also found in platelets (Marangos et al., 1980). Indeed, thrombocytes are now regarded as a component of the diffuse neuroendocrine system (Pearse, 1977). Blood platelets have been proposed as models for monoaminergic (Pletscher, 1978) and glutamatergic (Mangano and Schwarcz, 1981) neurones, and they also possess some gamma-aminobutyric acid (GABA)-ergic mechanisms (Hambley and Johnston, 1985): however, the detailed correspondence between platelet and CNS physiology remains to be firmly established in each case. It appears that the kinetics of human platelet serotonin (5-HT) uptake are similar to those seen in animal brain tissue (Stahl and Meltzer, 1978), but no human brain data are available for comparison. Moreover, thrombocytes are not thought to synthesize 5-HT from tryptophan (Stahl, 1985), and human platelet monoamine oxidase (MAO), 5-HT non-specific, has different properties from the multiple forms isolated from human brain (Youdim et al., 1972): therefore, human blood platelets can not be regarded as a good model for the 5-HT synthesis and turnover in human brain (Curzon, 1981). Nevertheless, it seems that platelet storage and release of 5-HT strictly resemble those observed in the CNS (Pletscher, 1988). Attempts to link changes in platelet 5-HT uptake with mental illness, behavioural abnormalities or neurological diseases have been made by several workers (s. Bongioanni, 1987, 1988). In the last few years many studies on the binding of tritiated imipramine (IMI) to human blood platelets, and its correlation with 5-HT uptake sites were reported (s. Rotman, 1983). IMI receptor shows a specific interaction with tricyclic antidepressants, as does the equivalent site on rat cerebral cortex synaptic membranes (Stahl, 1985). In cases where re-uptake into perisynaptic cells is the main route of disposal of a synaptically released neurotransmitter, the study of platelet uptake mechanisms provides an indirect method for studying dysfunctions of the analogous process in the CNS. Changes in platelet MAO activity have been found in several neuropsychiatric disorders (s. Bongioanni, 1987, 1988), although not always conclusive results have been gained, particularly as far as mental illness is concerned (Bongioanni, 1989).

Platelet aggregation, shape change and release reaction may serve as tools in neuropsychopharmacological research for testing the effects of various drugs, such as neuroleptics or antidepressants. For instance, recently, platelet aggregation has been used by Hefez et al. (1980) as an index for chlorpromazine efficacy, or by Youdim and Oppenheim (1981) as a tool for studying indirectly the properties of tryptolines in the CNS. Also the platelet shape change served to evaluate and quantitate the potencies of different 5-HT antagonists (s. Rotman, 1983). Moreover, Laubscher et al. (1979) studied the shape change induced by myelin basic protein, polyornithine, polylysine and protamine, which are known to cause neural depolarization in the CNS; Gudat et al. (1981) reported that nerve growth factor, substance P and thymopoietin caused the rapid onset of a shape change reaction, mostly by interacting with specific platelet membrane receptors. The platelet release reaction shares common features with the neuronal release of neurotransmitters. In both systems one can provoke a secretion via exocytosis with the plasma membrane or intracellular release by reserpine-like drugs, but the physiological stimulation for release in the neuron (electrical stimulation) is different from that of the platelet (thrombin, ADP).

The demonstration of alpha-adrenergic receptors (and a limited number of beta-adrenergic binding sites) on platelet plasma membrane (Boullin and Elliott, 1979; Steer and Atlas, 1982) has given the opportunity to study the alterations of some adrenergic receptor binding parameters and c-AMP production in platelets and, possibly, the changes of CNS adrenergic transmission in several neuropsychiatric conditions. Moreover, many adrenergic psychoactive drugs may be tested via platelet adrenergic receptors.

It is quite clear now that human blood platelets have at least two potential applications in neuropsychiatric research, the former as a model of nerve cell [neurone and gliocyte, too: the pharmacological characteristics of the rat platelet uptake site for GABA are similar to those of the GABA carrier found in CNS glial cells, and unlike those of neuronal high affinity uptake (Hambley and Johnston, 1985); moreover, the benzodiazepine (BDZ) binding site on human and rat platelet membrane is of the "peripheral" type, which is also found on gliocytes, rather than the GABA-receptor-linked "central" site found on neurones (Wang et al., 1980; Moingeon et al., 1984); in contrast, the neuronally located GABA transaminase is very similar in human platelets and brain tissue (White, 1979)]; the latter as a biochemical tool for drug testing. The use of platelets as a system of preliminary evaluation of the properties of newly synthesized polycyclic antidepressants was reported by Asscher (1981); more recently, Paul et al. (1982) have developed a radio-receptor assay for measuring the tricyclic antidepressant plasma levels, based on the competition of plasmatic tricyclic antidepressants with tritiated IMI for platelet binding sites.

In the present study human blood platelets have been utilized to investigate the effects of various BDZ on 3H-PK 11195 binding in normal and in demented subjects.

Materials and methods

Subjects

Thirty-five patients (mean age ± SD: 67.3 ± 7.6 years) — 16 males and 19 females (mean age ± SD: 64.5 ± 6.1 and 69.6 ± 8.8 years, respectively) — with senile dementia of Alzheimer type (SDAT), according to National Institute of Neurologic and Communicative Disorders and Stroke (NINCDS) criteria (1988), were studied. Clinical staging was defined according to Clinical Dementia Rating (CDR) scale by Hughes et al. (1982). By means of the Hamilton depression rating scale (HDRS) patients suffering from depression were excluded from the study.

The control group was formed by 30 healthy volunteers (mean age ± SD: 45.8 ± 6.2 years) — 12 males and 18 females (mean age ± SD: 43.1 ± 5.6 and 47.7 ± 7.3 years, respectively) — with no personal or family history of neuropsychiatric disorders.

Preparations of platelet membranes

Blood samples (30 ml) were obtained by venipuncture from patients and controls in the fasting state on the same day, between 8:00 and 9:00 a.m., collected into plastic tube containing 1 ml of 5% ethylendiamine tetra-acetic acid (EDTA), and spun at 200 × g for 15 min at 25°C. Platelet-rich plasma (PRP) was collected and spun at 4,000 × g for 20 min at 25°C. Platelet pellets were frozen at −80°C until assay, when they were homogenized (to obtain platelet lysates) in 20 ml of 50 mM Tris-HCl buffer, pH 7.4, at 4°C with a Brinkman polytron for 20 s, and centrifuged at 48,000 × g for 20 min. The pellets obtained were washed twice and finally resuspended in 20 ml of 50 mM Tris-HCl buffer, pH 7.4, at 4°C, and used for protein determination (Lowry et al., 1951) and for binding studies.

3H-PK 11195 binding

Four hundred microliters of platelet membranes (in triplicate) in 50 mM Tris-HCl buffer, pH 7.4, were incubated for 1 hour at 4°C with 3H-PK 11195 (0.05 ml) in absence (for total binding) or presence (for non-specific binding) of 0.001 ml of unlabelled Ro 5-4864. The samples were then filtered under vacuum over Whatman GF/B filters and washed four times with 5 ml of 50 mM Tris-HCl buffer, pH 7.4. The filters were placed in vials containing a scintillation mixture, and counted for radioactivity by means of a liquid scintillation counter. Eight different concentrations of 3H-PK 11195 (0.4–8.0 nM) were used to determine the platelet binding in each subject, and the data were evaluated by Scatchard analysis to obtain the maximal number of binding sites (B_{max}, expressed in fmols/mg of platelet proteins) and their affinity to the radioligand, i.e. the equilibrium dissociation constant (K_d, expressed in nM). Several BDZ were tested as inhibitors of specific 3H-PK 11195 binding to human blood platelet membranes: Ro 15-1788, Ro 5-4864, PK 11195, diazepam, oxazepam, flunitrazepam, clonazepam were added at various concentrations (from 0.1 to 10,000 nM) to the incubation mixture (containing 2.1 nM 3H-PK 11195). The drug concentrations able to inhibit 50% of 3H-PK 11195 binding (IC_{50}) were calculated from 3 different experiments.

Statistical analysis of the data for intergroup variations was performed using Student's t-test. All results were expressed as mean ± SD.

Results

A significant ($p < 0.01$) reduction of 21.7% in B_{max} average value was observed in platelet membranes of SDAT patients in respect with those of healthy controls; if males and females were considered separately, it was found that among females the difference between patients and controls was higher than that noted for male subjects (24.8% $p < 0.01$ vs 17.5% $p < 0.05$).

Table 1. Average B_{max} and K_d values of peripheral-type benzodiazepine binding sites in platelets of patients and controls

Subjects	N	B_{max} (fmols/mg protein)	K_d (nM)
Controls	30	3,219 ± 815	3.9 ± 0.7
Male	12	3,170 ± 822	4.0 ± 1.1
Female	18	3,251 ± 916	3.8 ± 0.9
SDAT patients	35	2,522 ± 793*	3.5 ± 1.1
Male	16	2,615 ± 771**	3.8 ± 1.3
Female	19	2,444 ± 839***	3.3 ± 0.9

All values are represented as mean ± SD
* significantly lower than controls ($p < 0.01$)
** significantly lower than male controls ($p < 0.05$)
*** significantly lower than female controls ($p < 0.01$)

The K_d values were similar in all subject groups (Table 1). The Scatchard plot in patients and controls was linear (r = 0.93–0.97), which suggests the presence of a single population of binding sites. The pharmacological profile of 3H-PK 11195 binding sites, assessed using different BDZ drugs (Table 2), shows that they are of "peripheral" type, because of the high affinity of Ro 5-4864 and the extremely low affinity of Ro 15-1788. No relevant differences have been found between patients and controls.

Discussion

In the present work it has been shown that human blood platelets possess only "peripheral" binding sites for BDZ with pharmacological properties similar to those of "peripheral" sites in brain (Marangos et al., 1982). If nowadays there is good evidence to suggest that "central" BDZ receptors mediate the pharmacological actions of the BDZ, and participate in the physiological expression of anxiety, seizures and sleep (Braestrup et al., 1982; Ninan et al., 1982; Mendelson et al., 1983), the precise pharmacological and physiological functions of the "peripheral" binding sites for BDZ in brain are unknown.

The data of a reduced number of BDZ binding sites (without any functional alterations, as shown by an unchanged average K_d value) on platelets of demented patients in comparison with that of healthy controls, might correlate with the neurodegenerative processes observed in SDAT: it could be supposed that either some cerebral biochemical changes concerning the BDZ binding sites are reflected in peripheral tissues, or certain systemic biochemical alterations coexist with central neurotransmitter abnormalities.

Table 2. Inhibition of specific 3H-PK 11195 binding to human platelet membranes by different benzodiazepines in patients and controls

Compound	IC_{50} (nM)	
	Controls	Patients
PK 11195	2.56	2.39
Ro 5-4864	7.62	8.71
Diazepam	96.18	94.27
Flunitrazepam	4,122.30	5,395.68
Clonazepam	4,761.54	5,131.86
Oxazepam	6,210.77	5,957.90
Ro 15-1788	>10,000*	>10,000*

Each IC_{50} value results from 3 different experiments.
The concentration of 3H-PK 11195 was 2.1 nM.
* this drug inhibited the binding only by 30–45% at the concentration of 10,000 nM

However, the human blood platelet does not serve only as a model of nerve cell, but also for testing the effects of several drugs: for instance, newly synthesized BDZ may be evaluated "in vitro" on platelet binding sites, thus getting useful information about their activity on "peripheral" type BDZ receptors in the CNS.

References

Asscher Y, Avnir D, Rotman A, Agranat I (1981) Active conformation of polycyclic antidepressants. J Pharmacol Sci 71: 122–124

Bongioanni P (1987) Platelets as a model in neurological studies: biochemical and clinical features. Arch Psicol Neurol Psichiat 48: 169–192

Bongioanni P (1988) Platelets as a model in psychiatric studies. NPS 8: 582–618

Bongioanni P (1989) Platelets and schizophrenia: platelet monoamine oxidase activity in schizophrenics. Riv Psichiat 24: 37–47

Boullin DJ, Elliott JM (1979) Binding of 3H-dihydroergocryptine to alpha-adrenoceptors of intact human platelets. Br J Pharmacol Chemother 66: 89P

Braestrup C, Schmiechen R, Neef G, Nielsen M, Petersen E (1982) Interaction of convulsive ligands with benzodiazepine receptors. Science 216: 1241–1243

Curzon G (1981) The turnover of 5-hydroxytryptamine. In: Pycock CJ, Taberner PV (eds) Central neutransmitter turnover. Croom Helm, London, pp 59–90

Gudat F, Laubscher A, Otten U, Pletscher A (1981) Shape change induced by biologically active peptides and nerve growth factors in blood platelets of rabbits. Br J Pharmacol 74: 533–538

Hambley JW, Johnston GAR (1985) Uptake of gamma-aminobutyric acid by human blood platelets: comparison with CNS uptake. Life Sci 36: 2053–2058

Hefez A, Oppenheim B, Youdim MBH (1980) Human platelet aggregation response to serotonin as an index of efficacy of chlorpromazine. In: Usdin E, Sourkes TL, Youdim MBH (eds) Enzymes and neurotransmitters in mental disease. J Wiley, Chichester, pp 76–93

Hughes CP, Berg L, Danziger WL, Cohen LA, Martin RL (1982) A new clinical scale for the staging of dementia. Br J Psychiatry 140: 566–572

Laubscher A, Pletscher A, Honegger CG, Richards JG (1979) Shape change of blood platelets brought about by myelin basic protein and other basic polypeptides. Arch Pharmacol 310: 87–92

Lowry OH, Rosebrough NJ, Farr AL, Randall RJ (1951) Protein measurement with the folin phenol reagent. J Biol Chem 193: 265–275

Mangano RM, Schwarcz R (1981) The human platelet as a model for the glutamatergic neuron: platelet uptake of L-glutamate. J Neurochem 36: 1067–1076

Marangos PJ, Campbell IC, Schmechel DE, Murphy DL, Goodwin FK (1980) Blood platelets contain a neuron-specific enolase. J Neurochem 34: 1254–1258

Marangos PJ, Patel J, Boulenger J-P, Clark-Rosenberg R (1982) Characterization of peripheral-type benzodiazepine binding sites in brain using 3H-Ro 5-4864. Mol Pharmacol 22: 26–32

Mendelson W, Cain M, Cook J, Paul S, Skolnick P (1983) A benzodiazepine receptor antagonist decreases sleep and reverses the hypnotic actions of flurazepam. Science 219: 414–416

Moingeon P, Dessaux JJ, Fellous R, Alberici GF, Bidart JM, Motté P, Bohuon C (1984) Benzodiazepine receptors on human blood platelets. Life Sci 35: 2003–2009

Ninan P, Insel T, Cohen R, Cook J, Skolnick P, Paul S (1982) Benzodiazepine receptor-mediated experimental "anxiety" in primates. Science 218: 1332–1334

Paul SM, Rehavi M, Hulihan B, Skolnick P, Goodwin FK (1982) A rapid and sensitive radio-receptor assay for tertiary amine tricyclic antidepressants. Comm Psychopharmacol 22: 56–59

Pearse AGE (1877) The diffuse neuroendocrine system and the APUD concept: related "endocrine" peptides in brain, intestine, pituitary, placenta, and anuran cutaneous glands. Med Biol 55: 115–125

Pletscher A (1978) Platelets as models for monoaminergic neurons In: Youdim MBH (ed) Essays in neurochemistry and neuropharmacology, vol 3. John Wiley, London, pp 49–101

Pletscher A (1988) Platelets as models: use and limitations. Experientia 44: 152–155

Rotman A (1983) Blood platelets in psychopharmacological research. Prog Neuropsychopharmacol Biol Psychiatry: 135–151

Stahl SM (1985) Platelets as pharmacologic models for the receptors and biochemistry of monoaminergic neurons In: Longenecker GL (ed) The platelets: physiology and pharmacology. Academic Press, London, pp 307–340

Stahl SM, Meltzer HY (1978) A kinetic and pharmacologic analysis of 5-hydroxytryptamine transport by human platelets and platelet storage granules: comparisons with central serotonergic neurons. J Pharmacol Exp Ther 205: 118–132

Steer ML, Atlas DC (1982) Demonstration of human platelet beta-adrenergic receptors using 125 I-labeled cyanopindolol and 125 I-labeled hydroxybenzylpindolol. Biochim Biophys Acta 686: 240–247

Wang JKT, Taniguchi T, Spector S (1980) Properties of 3H-diazepam binding sites on rat blood platelets. Life Sci 27: 1881–1888

White HL (1979) 4-Aminobutyrate:2-oxoglutarate aminotransferase in blood platelets. Science 205: 696–698

Youdim MBH, Collins GGS, Sandler M, Jones AB, Pare CMB, Nicholson WJ (1972) Human brain monoamine oxidase: multiple forms and selective inhibitors. Nature 236: 225–226

Youdim MBH, Oppenheim B (1981) The effect of tryptolines (1,2,3,4-tetrahydro-beta-carbolines) on monoamine metabolism and the platelet aggregation in human platelets. Neuroscience 6: 801–810

Authors' address: Dr. P. Bongioanni, Scuola Superiore di Studi Universitari e di Perfezionamento — "S. Anna", Pisa, Italy

J Neural Transm (1991) [Suppl] 33: 73–80
© by Springer-Verlag 1991

Pattern electroretinogram and luminance electroretinogram in Alzheimer's disease

K. Strenn[1], **P. Dal-Bianco**[2], **H. Weghaupt**[1], **G. Koch**[2], **C. Vass**[1], and **I. Gottlob**[1]

[1]First University Eye Clinic, Vienna, and [2]University Clinic of Neurology, Vienna, Austria

Summary. Visual symptoms are often among the first complaints of patients suffering from Alzheimer's disease and several studies showed a delay in flash visual evoked potentials. Hinton et al. (1986) described optic nerve degenerations in patients with Alzheimer's disease and Sadun published a dropout of retinal ganglion cells that range from 30% to 60%. The reduction of neurotransmitters, especially of acetylcholin, found in the brain might also occure in the retina. Therefore we examined the retinal functions of patients suffering from Alzheimer's disease.

In eight patients the pattern-electroretinograms and the scotopic and photopic luminance-electroretinograms were recorded and compared to an age-matched control group. We could not find any abnormalities in the pattern- and the luminance electroretinograms of patients with Alzheimer's disease. Although cholinergic cells have been found in the retina, our results did not reveal an involvement of retinal functions in Morbus Alzheimer.

Introduction

Alzheimer's disease (AD) is a disability of CNS with typically presenile dementia.

The symptoms include memory loss and total cognitive degeneration accompanied by a characteristic loss of neurons in the brain.

Visual symptoms are often among the first complaints of people with AD. This was first attributed to cerebral lesions, but recently much work has been conducted to investigate the primary visual pathway.

In 1986 Hinton et al. found an optic nerve degeneration in a post mortem study and Sadun et al. published a dropout of retinal ganglion cells ranging from 30% to 60%. Wright et al. (1986) described delayed flash visual evoked potentials while the pattern reversal evoked potential was of normal latency.

Although the causative mechanism of AD is unclear, a deficit of acetylcholin and of other neurotransmitters (somatostatin, serontonin, dopamin etc.) found in the brain seems to play a role in the pathogenesis of AD.

Since this reduction might also appear in the retina and cholinergic cells have been described in the retina, it was our aim to investigate electrophysiogical retinal function of patients with AD.

Materials and methods

Subjects

We examined 8 patients with Alzheimer's disease, the diagnosis based on the DSM-III-criteria. The age ranges from 52 to 80 years, the mean age was 63.6 years. Their mean Snellen acuity was 0.8 for the right eyes, 0.7 for the left eyes. The patients were not highly demented and still cooperative. An age-matched group of ten volunteers, mean age from 49 to 82 years, served as the control.

Methods

Luminance-ERG

For recording, a Nicolet Ganzfeld System was used. The duration of one stimulus was 100 ms. The L-ERG was amplified and averaged with a Nicolet CA 2000. Henkes corneal-ERG electrodes were used and the reference electrode was placed at the ipsilateral temple.

The pupils were dilated using Tropicamid eye drops. The scotopic ERG was elicited with four single blue test flashes with increasing intensities (0.01, 0.02, 0.024, 0.075 lux. sec.) after 25 minutes of dark adaption. In the presence of a blue adaption light (Kodak, Wratten 47), the photopic ERG (a- and b-wave) was recorded with red/yellow flashes (Wratten filter 12/25) at four intensities (0.024, 0.032, 0.04, 0.074 lux. sec.) and was averaged 8 times.

Pattern reversal-ERG

Pattern ERG was recorded using a gold foil electrode, placed in the lower fornix. The reference electrode (silver-silverchlorid) was placed to the ipsilateral temple. Pupils were not dilated and the best optical correction was given. For stimulation a checkerboard pattern was produced on a monitor at a distance of 1.5 meters.

The test field was 18*14 deg. arc. Each checkerboard subtended 65 min of arc, the mean luminance was 30 cd/m2. The pattern was reversed at a frequency of 4 rev. per second. The contrast was 97%.

280 sweeps were averaged with a Nicolet CA 2000.

The amplitude was measured between the first negative and the first positive peak. The latency is defined between the onset of the pattern reversal and the first positive peak.

Results

Luminance-ERG

The mean values and standard deviation of the scotopic b-wave latencies (Fig. 3) and amplitudes (Fig. 4) of all patients and control subjects are presented.

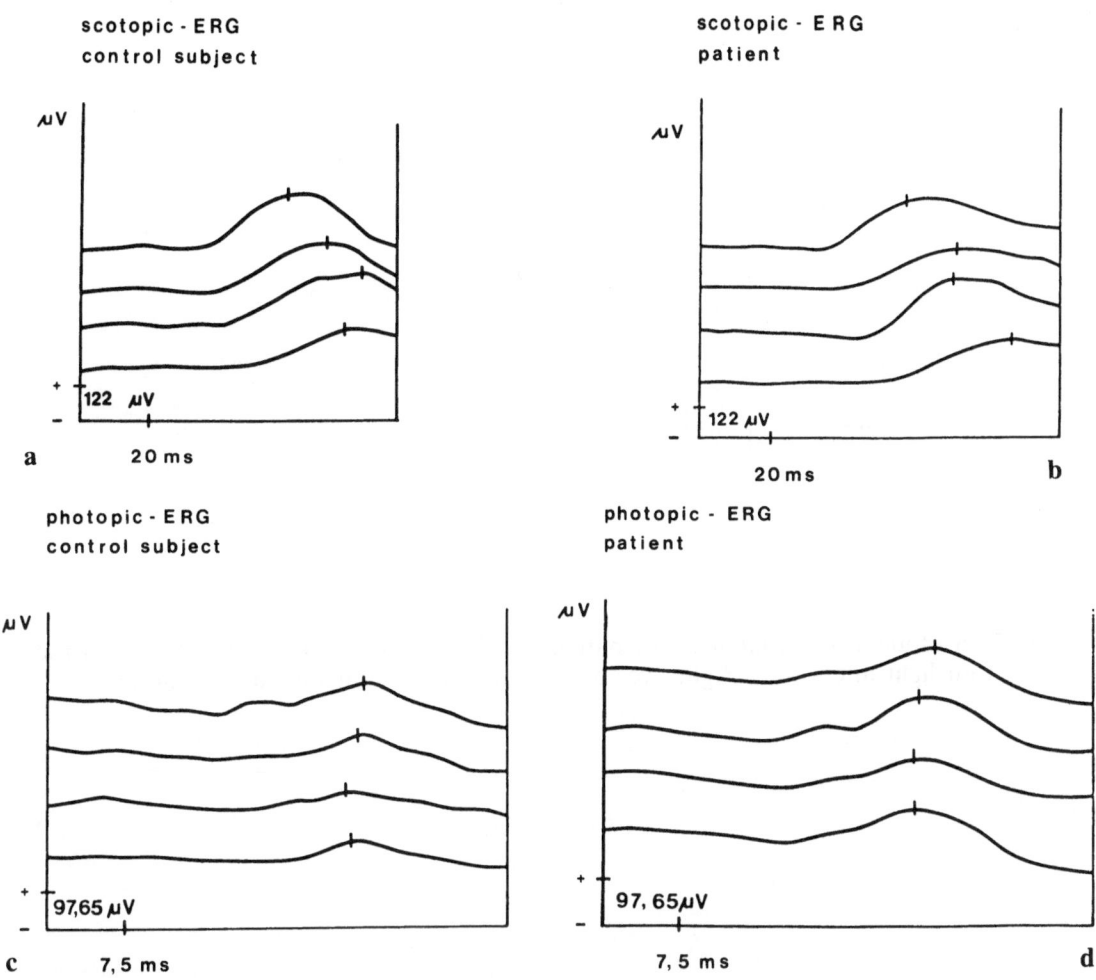

Fig. 1a. Original scotopic-luminance FRGs of one control subject and **b** one patient with AD. **c** Original photopic-luminance ERGs of one control subject and **d** of one patient with AD

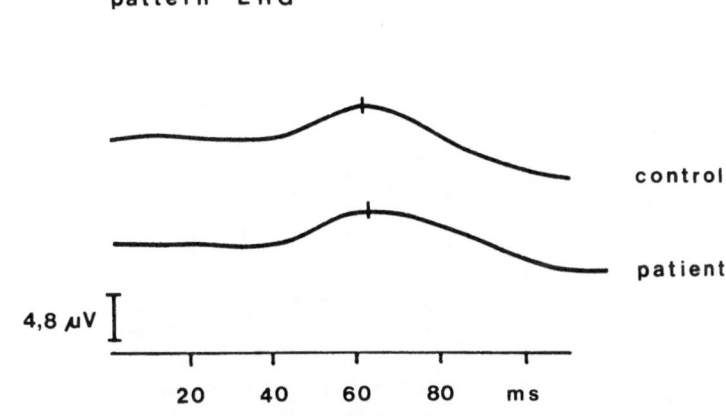

Fig. 2. Original pattern-ERGs of one patient with AD (left position) and one control subject (right position)

scotopic b - latency

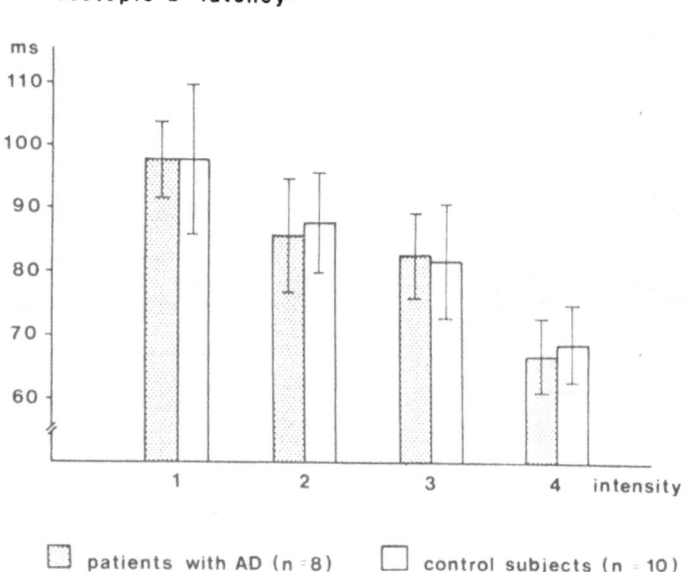

Fig. 3. Scotopic b-wave latencies of patients with AD (n = 8) and controls (n = 10) at four light intensities. Right eyes only. Mean values and standard deviations

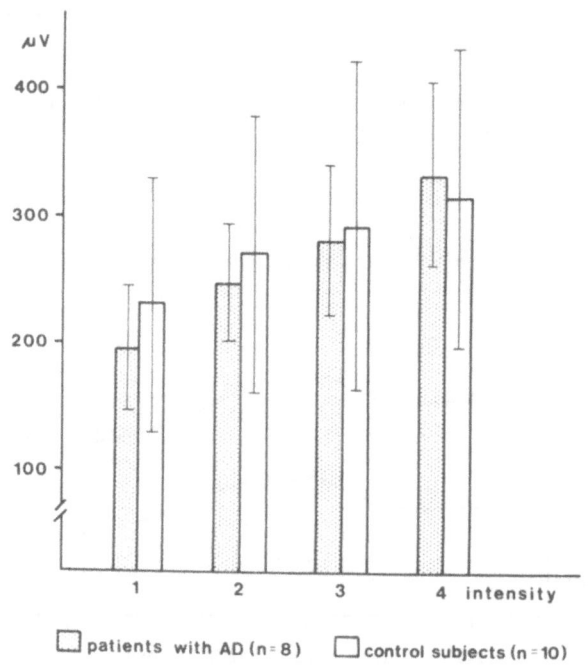

Fig. 4. Scotopic b-wave amplitudes of patients with AD (n = 8) and controls (n = 10) at four light intensities. Right eyes only. Mean values and standard deviations

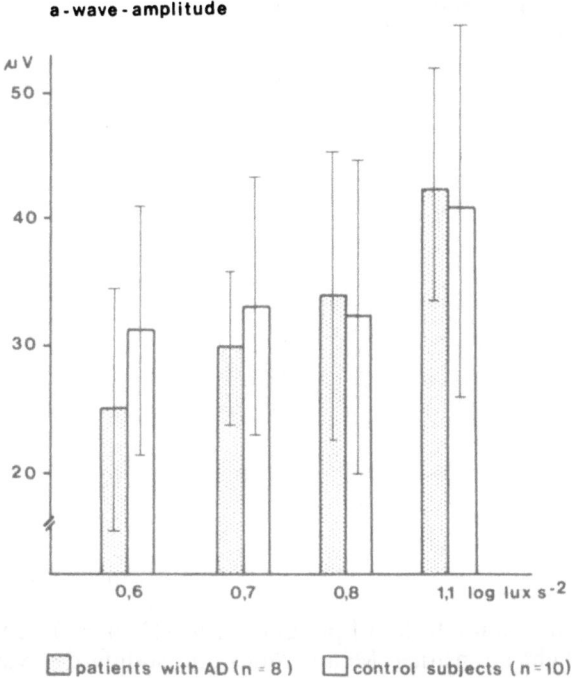

Fig. 5. Photopic b-wave amplitudes of patients with AD (n = 8) and controls (n = 10) at four light intensities. Right eyes only. Mean values and standard deviations

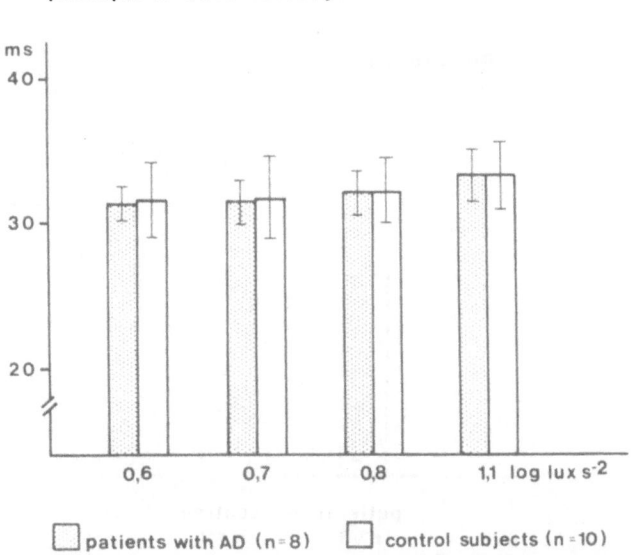

Fig. 6. Photopic b-wave latencies of patients with AD (n = 8) and controls (n = 10) at four light intensities. Mean values and standard deviations

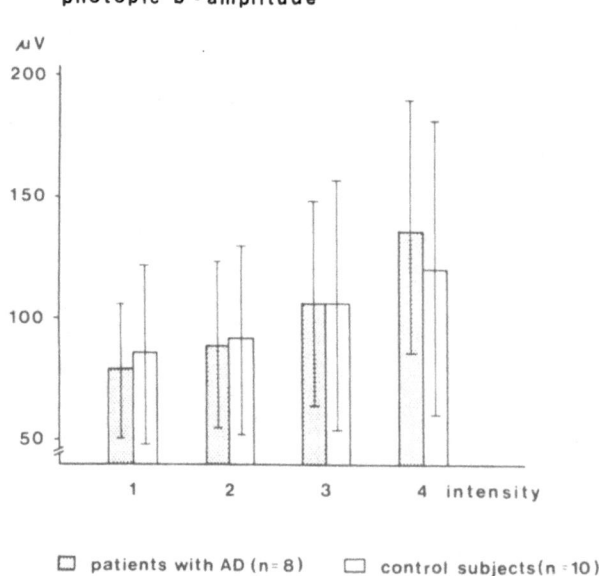

Fig. 7. Photopic a-wave amplitudes of patients with AD (n = 8) and controls (n = 10) at four light intensities. Mean values and standard deviations

Subjects did not reveal significant differences.

Like in the scotopic ERG, we could not detect any abnormalities in the photopic ERG of the AD patients.

As seen in Figs 5, 6, and 7, the patients photopic a-waves and b-waves remained within standard values at all intensities in latency as well as in amplitude.

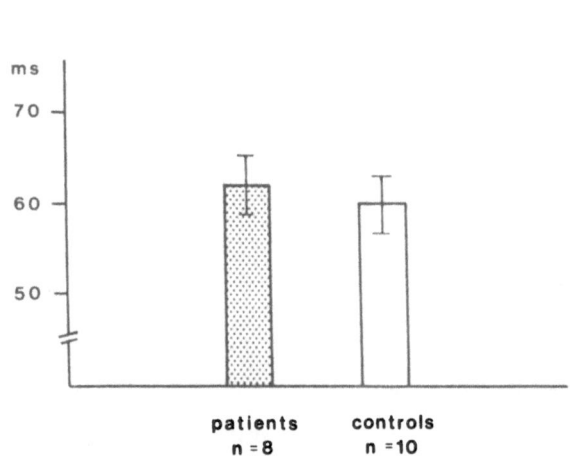

Fig. 8. Pattern-ERG latencies of patients with AD (n = 8) and controls (n = 10). Mean values and standard deviations

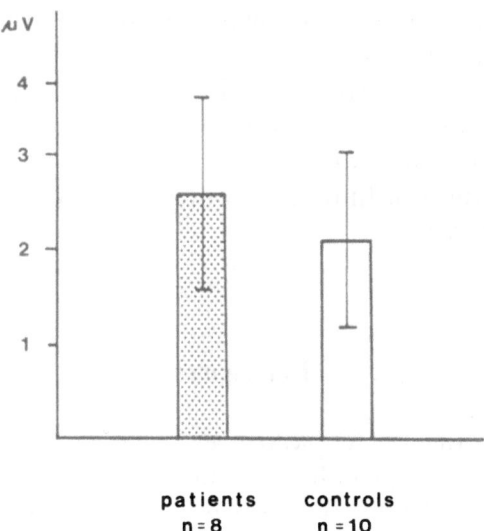

Fig. 9. Pattern-ERG amplitudes of patients with AD (n = 8) and controls (n = 10). Mean values and standard deviations

Pattern evoked potential

In Figs. 8 and 9 the mean values and the standard deviations of the pattern-ERGs are shown. For both figures the left column represents the mean value recorded from the patients' right eyes. The right column gives the mean value of the control group.

We could not detect any significant difference between both groups.

Example of the original curves of the luminance-ERGs and the pattern-ERGs recorded from one patient and one control subject are presented in Figs. 1 and 2.

Discussion

The etiology and pathogenesis of Alzheimer's disease are poorly understood. The causative mechanism proposed include virus infection, immunological incompetence, aluminium toxicity, vascular dysfunction and enzym deficiency.

Post-mortem neurochemical studies showed deficits in several neurotransmitters. The outstanding reduction of acetylcholin in brain, is proposed to cause several cerebral symptoms characteristic for AD.

Since cholinergic amacrine cells have been described in the retina, the reduction of acetylcholin might have an influence on the retinal function. Our result of a normal luminance electroretinogram of patients with AD, compared to an age-matched control group, does not point to a loss of retinal function.

Hinton et al. (1986) found in a post mortem study an optic nerve degeneration together with a reduction of retinal ganglion cells in patients having suffered from AD. Sadun et al. published a dropout of retinal ganglion cells ranging from 30% to 60% (1986).

Supposing that the pattern reversal electroretinogram reflects the function of the ganglion cell layer, our result of a normal pattern-ERG does not reveal any loss of function there.

In summary, we did not find an involment of retinal functions in patients with Alzheimer's disease.

References

Hinton D, Sadun A, Blanks JC, Miller C (1986) Optic nerve degeneration in Alzheimer's disease. N Engl J Med 315: 485–487

Hutton T, Albrecht JW, Sharpiro I, Johnston C (1987) Visual information processing and dementia. Neuroophtalmology 7(2): 105–112

Orwin A, Wright CE, Harding GF, Rowan DC, Rolfe EB (1986) Serial visual evoked potential recording in Alzheimer's disease. Br Med J: 293

Rossor MN (1982) Dementia. Lancet: 1200–1203

Rossor M (1987) Cerebral aging. Eye 1: 171–174

Sadun A, Borchert M, DeVita E, Hinton D, Bassi CJ (1987) Assessment of visual impairment in patients with Alzheimer's disease. Am J Ophtalmol 104: 113–120

Visser SL, Stam FC, Van Tilburg W, op den Velde W, Blom JL, de Rijke W (1976) Visual evoked response in senile and presenile dementia. Electroencephalogr Clin Neurophysiol 40: 385–392

de Vries-Khoe LH, Spekreijse H (1982) Maturation of luminance and pattern EPs in man. Doc Ophtalmol Proc Series 31: 461–475

Wright CE, Harding GFA, Orwin A (1984) Presenile dementia-the use of the flash and pattern VEP in diagnosis. Electroencephalogr Clin Neurophysiol 57: 405–415

Wright CE, Williams DE, Draso N, Harding GFA (1985) The influence of age on the electroretinogram and visual evoked potential. Doc Ophtalmol 59: 365–384

Wright CE, Harding GFA, Orwin A (1986) The flash and pattern VEP as a diagnostic indicator of dementia. Doc Ophtalmol 62: 89–96

Authors' address: Dr. K. Strenn, Krankenhaus Scheibbs, Eisenwurzenstrasse 26, A-3270 Scheibbs, Austria

J Neural Transm (1991) [Suppl] 33: 81–92
© by Springer-Verlag 1991

Clinical comparison of dementia in Parkinson's and Alzheimer's disease

I. Marschall, P. Dal-Bianco, J. Maly, E. Auff, J. Hufgard, M. Mraz, and **L. Deecke**

Neurological University Clinic, Vienna, Austria

Summary. Neuropsychological, neuropathological and neurochemical findings show different types of dementias. Few of them have been able to confirm a division into "subcortical" and "cortical" dementia, so this concept has to be questioned. The present clinical study compared type and severity of dementia in 12 Parkinson-patients (PD) and 12 Alzheimer-patients (AD). The age-adjusted normal value differed a significantly from both patient groups. No significant difference in pattern of neuropsychological deficits between PD- and AD-patients was apparent. However, after similar duration of illness, dementia was more severe in AD- than in PD-patients.

Introduction

The term "subcortical dementia", first mentioned about fifteen years ago (Albert, 1974), describes a clinical syndrome of cognitive impairment, associated with progressive supranuclear palsy, normal pressure hydrocephalus, and Huntington's disease. Typical symptoms are deterioration in memory and learning, slowness of intellectual function and apathy but no difficulties in language, apraxia, and perception — symptoms that were considered to be more characteristic of cortical dementias such as Alzheimer's disease. Cognitive impairment in depression was also classified as subcortical dementia (Birkmayer and Riederer, 1987; Hart and Kwentus, 1987; Lees, 1989; Rogers et al., 1987). The advantage of such a division into subcortical and cortical dementia (Mayeux et al., 1983, 1981) has been questioned in recent publications and clinical, neuropathological and neurochemical studies show little evidence to support this division. The aim of this study is the clinical comparison of type and severity of dementia in Parkinson's (PD) and Alzheimer's disease (AD).

Table 1. Demographic and anamnestic data

n Age (a)	PD 12			AD 12		
	median	min	max	median	min	max
	74,5	62	87	63,5	53	74
Therapy	L-Dopa			"nootropica"		
Duration of illness (a)	14,5	5	24	8	2	14
Severity	41,5	22	61	6/I	3/III	3/IV
	Columbia rating scale			Summer's Score		

Patients and methods

12 PD-patients aged from 62 to 87 years and 12 AD-patients between 53 and 74 years were examined. The duration of illness in PD-patients ranged from 5 to 24 years, in AD-patients from 2 to 14 years. The range of severity in PD [Columbia Rating Scale (Duvoisin, 1970)] was between 22 and 61 points, AD-patients scored at stage I to IV [Summer's Score (Summers and Viesselman, 1981)]. PD-patients were receiving L-dopa-treatment, AD-patients "nootropica" (Table 1).

All patients underwent a clinical neurological examination. In PD-patients rigidity and bradykinesia dominated. For exclusion of other etiologies of dementia all patients had a CAT-scan (only diffuse ventricular and subarachnoidal enlargement). The neuro-

Table 2. Neuropsychological tests

1. Psychophysiological measurement:
   Vigilance
   Concentration
   Reaction time measurement
   Perception
   Motor performance

2. Objective measurement:
   Language ability (AAT)
   Mental functions (Luria)
   Intelligence
   Memory and concentration

3. Subjective measurement:
   Rating scales

4. Projective measurement:
   Rorschach test

The neuropsychological test battery includes MMSE and four measurements: verbal and nonverbal memory, language, motor and attention performances were examined

Table 3. Comparison of total scores

	Score	SD		
PD	82	46	n.s.	
AD	62	41		s.
Controls	249	37	s.	

Comparison of total scores of the neuropsychological test performances: there is a significant difference between the age — adjusted standard value and both patient groups. The total scores did not differ significantly between PD- and AD-patients

psychological test battery included MMSE, verbal and nonverbal memory, language, motor, and attention performances (Table 2).

Results

Compared to the aged-adjusted standard value of 249 points (SD 37) in the neuropsychological test battery PD-patients reached an average of 82 points (SD 46) and AD-patients 62 points (SD 41). In Table 3 the total scores of the neuropsychological test performances are compared: there is a significant difference between the aged-adjusted standard value and both patient groups. The total scores did not differ significantly between PD- and AD-patients. Only the MMSE-score and the labyrinth test differed significantly between the patient groups (Table 4). No significant difference was found between the two patient groups for the other test results. After similar duration of illness, dementia was more severe in AD-patients than in PD-patients.

The parameter of precise movements which includes the independent factors "movement disorder and accuracy of sighting" (III-MLS) as well as "speed of precise arm-hand-motions" (IV-MLS) is measured in the labyrinth test. In our study both precision and speed of a complex motion are disturbed significantly in the AD-patient group. Apraxia is possibly the cause of this result.

Table 4. Significant different parameters

	PD	AD	p
MMSE	20,333	11,667	≦0,01
Labyrinth test	45,833	119,25	≦0,02

Only the MMSE-score and the labyrinth test differed significantly between the patient groups. No significant difference was found between the two patient groups for the other test results

Discussion

Clinical studies

In the present study no significant difference was found in the neuropsychological pattern of PD- and AD-dementia. Similar profiles of dementia in both diseases, based on a combination of subcortical and cortical degeneration, have also been described by Mayeux et al. (1983). Although the clinical diagnosis of dementia is rarely able to predict the etiology (Boller et al., 1989), Hakim and Mathieson (1978) described a 56% risk of dementia in PD-patients. Leverenz and Sumi (1986) found a prevalence of 50% of extrapyramidal disorders in patients with Alzheimer's disease. The fact that PD- and AD-patients show similar neuropsychological, neuropathological and neurochemical changes (Boller et al., 1980; Candy et al., 1983; Rinne et al., 1984; Whitehouse, 1986, 1989; Whitehouse et al., 1987, 1988; Whitehouse and Unnerstall, 1988) opposes the division of dementia (Boller et al., 1980; Mayeux et al., 1983; Whitehouse, 1986; Whitehouse and Unnerstall, 1988) into the subcortical and cortical type. Albert et al. (1974) proposed the concept of "subcortical dementia" based on the hypothesis of impaired function of the reticular activating systems and/or a disconnection of reticular activating systems from thalamic and subthalamic nuclei. Sweet et al. (1976) and Halgin et al. (1977) described only initial improvement of cognitive performances in PD-patients following L-dopa-treatment. In 1986 Huber et al. defending the concept of "subcortical and cortical dementia" examined PD- and AD-patients. A neuropsychological procedure specifically designed for cortical disorders was used. Neither onset, stage nor duration of disease could be gathered from this study. According to Huber the results differentiated between dementia syndromes, and the pattern of performance was consistent with the subcortical-cortical hypothesis. In spite of clinical and apparative investigations the diagnosis of Alzheimer's disease — first documented by A. Alzheimer 1901 (Hoff and Hippius, 1989) — can only be confirmed neuropathologically (Boller et al., 1989; Haxby et al., 1986; Lauter et al., 1986). A lot of teams are working on the neuropathological changes and clinical profile of this system disorder (Freedman and Oscar-Berman, 1986; Haxby et al., 1986; Leverenz and Sumi, 1986; Mayeux et al., 1985; Pearce, 1974; Rapcsak et al., 1989). With regard to dementia, research on extrapyramidal disorders — especially Parkinson's disease (Flowers et al., 1984; Heston, 1980; Lees and Smith, 1983; Levin et al., 1989; Liebermann et al., 1979; Loranger et al., 1972; Martin et al., 1973; Mortimer et al., 1982; Pollock and Hornabrook, 1966; Rafal et al., 1984; Starkstein et al., 1987; Talland, 1962) — has been undertaken. The clinical distinction of these dementias even led to the etiology question (Albert et al., 1974; Boller et al., 1980; Freedman and Oscar-Berman, 1986; Huber et al., 1986; Mayeux et al., 1983; Shankar et al., 1989). A lot of neuropathological and neurochemical studies (Shankar et al., 1989) searched for connexions and differences between the dementias of Parkinson's and Alzheimer's disease (Monte et al., 1989; Whitehouse, 1986; Whitehouse et al., 1988; Zweig et al., 1989). All attempts to correlate quantity of morphological changes and severity of dementia were in vain (Earnest et al., 1979; Sroka et al., 1981). There are extending problems in the diagnosis of dementia in the aged due to confuse diagnostic criteria and terminology. DSM-III criteria are mainly used to classify dementias (Denzler et al., 1986; Hoff and Hippius, 1989). A lot of psychometric tests are available for clinical characterization of different types of dementias — e.g. WAIS (Wechsler Adult Intelligence Scale), SPMSQ (Short Portable Mental Status Questionnaire), NART (New Adult Reading Test), MSQ (Mental Status Questionnaire), DSM (Diagnostic and Statistical Manual of Mental Disorders), MOMSSE (Mattis Organic Mental Syndrome Screening Examination), MMSE (Mini-Mental-State-Examination) (Dick et al., 1984), Kew-test (Hare, 1978). These short tests define pathological cognitive and aged-associated memory impairment irrelevant of etiology. Because of the variety of screenings it is almost impossible to compare the results. A constant problem with any detailed test is the relatively long time required for testing (30 minutes). This often exceeds the capacity of even mildly impaired patients. Reaction time [PD, (Evarts et al., 1981; Rogers et al., 1987)], motivation, emotional stability are additional variables causing inter- as well as intra-individual results and bring about a wide scattering of test results. A pseudo-dementia test result can be achieved by a temporary depressive mood or depression. PD and depression are combined in 90% according to Mindham et al. (1982), 37–90% Mayeux et al. (1981) and about 9% Piccirilli et al. (1984). Dementia in PD occurs in our study in 40%, in 40% according to Cummings (1988), 20–40% Mortimer et al. (1985), 30% Mindham (1970), 25–80% Drachman and Stahl (1975), 56% Hakim and Mathieson (1978, 1979), 10–15% Brown et al. (1984), and 15% Lees (1985). Inclusion of symptoms for bradyphrenia (Lees, 1985) — lack of spontaneity, imagination and initiative, emotional instability, a tendency to repetition and some word-finding difficulties — by some authors, accounts for the relatively high percentage of dementia in patients with PD. "Bradyphrenia" and "subcortical dementia" have even been used synonymously in literature (Lees, 1989). AD-patients suffer from extrapyramidal features with progression of the disease in 40% (our study), in 85% Wetterling (1989), and 61% Drachman and Stahl (1975). J. Pearce (1974) found extrapyramidal signs in 40 out of 65 AD-patients and no improvement with levodopa treatment in 10% of the cases. This result suggests that the dopamine deficiency cannot be the only cause of parkinsonian symptoms (Pearce, 1974) in AD-patients. Neuropsychological deficits can precede motor symptoms in PD (Lees, 1989). With progression of the disease reduced verbal learning (Taylor et al., 1986) and visuospatial impairment (Boller et al., 1984; Sala et al., 1986; Taylor et al., 1986; Villardita et al., 1982) occur even in nondemented PD-patients. However verbal learning in demented PD-patients is not reduced to such an extent as in AD-patients (Pillon et al., 1986). Different neuropsychological deficits can occur — e.g. passive memory deficiency (Mayeux et al., 1981), psychomotor deceleration (Cummings, 1988; Taylor et al., 1986), and defective concept formation (Bowen et al., 1975; Caltagirone et al., 1989; Cools et al., 1984; Cummings, 1988; Flowers and Robertson, 1985; Taylor et al., 1986). Language aptitude does not normally show dysphasic elements (Cummings, 1988; Mortimer et al.,

1985; Villardita et al., 1982). These results agree with drug induced MPTP-parkinsonism where memory efficiency was hardly impaired (Stern and Langston, 1985). So called cortical deficits — e.g. constructional apraxia occuring in nearly all AD-patients (Rapcsak et al., 1989) — were also observed by Villardita et al. (1982) in PD-patients.

Essentially more PD- (Cummings, 1988; Mindham, 1970; Pillon et al., 1986; Villardita et al., 1982) than AD-patients (Pillon et al., 1986) suffer from depression. According to Oyebode et al. (1986) dementia of the Alzheimer type occurs in 7% of PD, and a further 65% of PD-patients show mild cognitive deterioration. Hietanen and Teräväinen (1984) found that verbal and visuoperceptive disturbance correlated directly with age, whereas memory impairment correlated directly with duration of disease. Mindham et al. (1982) and others (Matthews and Haaland, 1979; Mayeux et al., 1981; Oyebode et al., 1986) also found a positive correlation of duration of illness and severity of dementia. This correlation only applied to PD-patients with rigidity and slowness of movements (Mayeux et al., 1981; Mortimer et al., 1985). PD-patients with tremor — as in our study — seldomly showed distinct signs of memory impairment (Mortimer et al., 1982). According to Garron et al. (1972) onset of PD correlates with severity of intellectual impairment: the later the onset of disease, the more frequent the signs of cognitive detereoration. This applies for PD-patients with bradykinesia. Direnfeld et al. (1984) worked out following relationship of PD-laterality to neuropsychological deficit: more severe cognitive deterioration was found in PD-patients affected on the left. This could be due to functional or anatomical asymmetries of dopaminergic systems in the CNS. Not only duration and type of PD but especially long-term drug therapy (Portin and Rinne, 1984; Sweet et al., 1986) seem to influence intellectual efficiency. In a retrospective study only initial improvement of cognitive functions was shown by Sweet et al. (1976) and Halgin et al. (1977). Drug holidays reduced confused states following levodopa treatment. The effect of L-dopa on intellectual efficiency is not shown in this study (Sweet et al., 1976). According to Mohr et al. (1989), the ability to reproduce complex verbal material while receiving medication was enhanced. On the other hand Smet et al. (1982), showed that only vigilance and not cognitive efficiency increased. Reduction in plasma dopamine level between the time of original learning and subsequent memory retrieval cause a state-dependent memory impairment in PD according to Huber et al. (1989). Furthermore, demented PD-patients show a higher sensitivity to anticholinergic drugs (Smet et al., 1982) than non-demented patients. Long-term dopaminomimetic therapy can cause response fluctuations in motor function and to a lesser extent mood instability, whereas cognitive functions — especially language and memory are hardly affected (Brown and Marsden, 1984; Delis et al., 1982; Girotti et al., 1986).

Neuropathological studies

Both subcortical and cortical changes are typical of most dementias. Subcortical structures like nucleus basalis of Meynert, nucleus amygdalae, and raphe nuclei are affected in Alzheimer's disease (Bondareff et al., 1982; Candy et al., 1983; Leverenz and Sumi, 1986; Whitehouse, 1986; Whitehouse and Unnerstall, 1988; Whitehouse et al., 1981). Loss of neurons in nucleus basalis of Meynert and locus ceruleus is characteristic of PD (Candy et al., 1983; Cash et al., 1987; Gaspar and Gray, 1984; Whitehouse, 1986). Typical Alzheimer-like changes are found in both demented and nondemented PD-patients (Boller et al., 1980; Hakim and Mathieson, 1978, 1979; Whitehouse et al., 1983): according to Boller et al. (1980) they occur in 42% demented and 25% nondemented PD-patients. In addition atrophy and Lewy bodies are found in cortical layers of PD-patients. Hence so called "subcortical dementia" was explained on the hypothesis of cortical alterations (Earnest et al., 1979; Yoshimura, 1983). A specific clinical profile is presented by MPTP-induced parkinsonian syndrom lacking cortical and subcortical Lewy bodies (Lewin, 1985). Alzheimer-untypical amyloid plaques were found in a neuro-genetic disorder with dementia and parkinsonism described by Rosenberg et al. (1989). No alzheimer-like pathological features were found in five cases of demented patients by Heilig et al. (1985). Torack and Morris (1986) described a case of mesolimbocortical dementia with neuropsychological features exceeding those of PD and AD.

Neurochemical studies

PD and AD both share certain neuron and neurotransmitter system abnormalities (Bowen et al., 1977; Drachman and Leavitt, 1975; Eggertson and Sima, 1986; Javoy-Agid and Agid, 1980; Kish et al., 1985; Perry et al., 1985; Scatton et al., 1982; Whitehouse, 1986; Whitehouse et al., 1987; Whitehouse and Unnerstall, 1988; Wetterling, 1989). Degeneration of cholinergic neurons and decrease of choline acetyltransferase (CAT) occur in both disorders (Atack et al., 1988; Perry et al., 1983; Rinne et al., 1984; Whitehouse and Unnerstall, 1988; Whitehouse, 1989), although a recent study has shown a case of normal CAT-activity (Zweig et al., 1989). Some dysfunction of cholinergic neurons was found in mildly demented PD-patients and correlated positively to intellectual impairment (Whitehouse and Unnerstall, 1988; Whitehouse et al., 1983). In PD and progressive supranuclear palsy, dopaminergic changes are accompanied by subcortical and cortical degeneration but are only to some extent responsible for dementia (Pillon et al., 1989). Similar findings for noradrenergic and serotoninergic markers have been reported (Mann et al., 1983; Zweig et al., 1989). The results of CSF-studies are not uniform: nevertheless, they can contribute to the classification of dementia. PET-investigations (Phelps et al., 1986) can quantify in vivo regional receptor density and transmitter concentration in various systems thus contributing to the diagnosis of dementia.

Conclusion

It is evident from neuropsychological, neuropathological and neurochemical findings that different types of dementias exist. A precise clinical and neuropsychological characterization should be strived for, inspite of the many difficulties. Few clinical, neuropsychological, neuropathological and neurochemical findings have been able to confirm a division into "subcortical" and "cortical" dementia, so this concept has to be questioned. The present clinical study compared type and severity of dementia in 12 PD- and 12 AD-patients. The age-adjusted normal value differed significantly from both patient groups. No difference in pattern of neuropsychological deficits between PD- and AD-patients was apparent. However, after similar duration of illness, dementia was more severe in AD-patients than in PD-patients.

References

Albert ML, Feldmann RG, Willis AL (1974) The "subcortical dementia" of progressive supranuclear palsy. J Neurol Neurosurg Psychiatry 37: 121–130

Atack JR, May C, Kaye JA, et al (1988) Cerebrospinal fluid cholinesterases in aging and in dementia of the Alzheimer type. Ann Neurol 23: 161–167

Birkmayer W, Riederer P (1987) Die Parkinson-Krankheit/Biochemie, Klinik, Therapie, 2. Aufl. Springer, Wien New York

Boller F, Mizutani T, Roessmann O, Gambetti P (1980) Parkinson disease, dementia, and Alzheimer disease: clinicopathological relations. Ann Neurol 7: 329–335

Boller F, Passafinne D, Keefe NC, et al (1984) Visuospatial impairment in Parkinson's disease. Role of perceptual and motor factors. Arch Neurol 41: 485–490

Boller F, Lopez OL, Moosey J (1989) Diagnosis of dementia: clinicopathologic correlations. Neurology 39: 76–79

Bondareff W, Mountjoy ChQ, Roth M (1982) Loss of neurons of origin of the adrenergic projection to cerebral cortex (nucleus locus ceruleus) in senile dementia. Neurology 32: 164–168

Bowen FP, Kamienny RS, Burns MM, Yahr MD (1975) Parkinsonism: effects of levodopa treatment on concept formation. Neurology 25: 701–704

Bowen DM, Smith CB, White P, et al (1977) Chemical pathology of the organic dementias. Brain 100: 427–453

Brown RG, Marsden CD (1984) How common is dementia in Parkinson's disease? Lancet ii: 1262–1265

Brown RG, Marsden CD, Quinn N, Wyke MA (1984) Alterations in cognitive performance and affect-arousal state during fluctuations in motor function in Parkinson's disease. J Neurol Neurosurg Psychiatry 47: 454–465

Caltagirone C, Carlesimo A, Nocentini U, Vicari S (1989) Defective concept formation in Parkinsonians is independent from mental deterioration. J Neurol Neurosurg Psychiatry 52: 334–337

Candy JM, Perry RH, Perry EK, et al (1983) Pathological changes in the nucleus of Meynert in Alzheimer's and Parkinson's disease. J Neurol Sci 59: 277–289

Cash R, Dennis T, L'Henreux R, et al (1987) Parkinson's disease and dementia: norepinephrine and dopamin in locus ceruleus. Neurology 37: 42–46

Cools AR, Van Den Bercken JHL, Horstink MWI, et al (1984) Cognitive and motor shifting aptitude disorder in Parkinson's disease. J Neurol Neurosurg Psychiatry 47: 443–453

Cummings JL (1988) The dementia of Parkinson's disease: prevalence, characteristics, neurobiology, and comparison with dementia of the Alzheimer type. Eur Neurol 28: 15–23

Delis D, Direnfeld L, Alexander MP, Kaplan E (1982) Cognitive fluctuations associated with on-off phenomenon in Parkinson's disease. Neurology 32: 1049–1052

Denzler P, Kessler J, Markowitsch HJ (1986) Möglichkeiten und Mängel der psychometrischen Demenz-Diagnostik. Fortschr Neurol Psychiat 54: 382–392

Dick JPR, Guiloff RJ, Stewart A (1984) Mini mental state examination in neurological patients. J Neurol Neurosurg Psychiatry 47: 496–499

Direnfeld LK, Albert ML, Volicer L, et al (1984) Parkinson's disease. The possible relationship of laterality to dementia and neurochemical findings. Arch Neurol 41: 935–941

Drachman DA, Stahl S (1975) Extrapyramidal dementia and levodopa. Lancet i: 809

Drachman DA, Leavitt J (1975) Are neurotransmitter systems specific for cognitive functions? Neurology 25: 349

Duvoisin RC (1970) The evaluations of extrapyramidal disease. In: Ajuriaguerra dee, Gauthier G (eds) Monoamine, Noyeaux gris centraux et Syndrome de Parkinson. George and Cie, Geneva, pp 313–325

Earnest MP, Heaton RK, Wilkinson WE, Mauke WF (1979) Cortical atrophy, ventricular enlargement and intellectual impairment in the aged. Neurology 29: 1138–1143

Eggertson DE, Sima AAF (1986) Dementia with cerebral lewy bodies. A mesocortical dopaminergic defect. Arch Neurol 43: 524–527

Evarts EV, Teräväinen H, Calne DB (1981) Reaction time in Parkinson's disease. Brain 104: 167–186

Flowers KA, Pearce I, Pearce JMS (1984) Recognition memory in Parkinson's disease. J Neurol Neurosurg Psychiatry 47; 1174–1181

Flowers KA, Robertson C (1985) The effect of Parkinson's disease on the ability to maintain a mental set. J Neurol Neurosurg Psychiatry 48: 517–529

Freedman M, Oscar-Berman M (1986) Selective delayed response deficits in Parkinson's and Alzheimer's diseases. Arch Neurol 43: 886–890

Garron DA, Klawans HL, Norin F (1972) Intellectual functioning of persons with idiopathic Parkinsonism. J Nerv Ment Dis 154: 445–452

Gaspar P, Gray F (1984) Dementia in idiopathic Parkinson's disease. A neuropathological study. Acta Neuropathol 64: 43–52

Girott F, Carella F, Grassi MP, et al (1986) Motor and cognitive performances of Parkinsonian patients in the on and off phases of the disease. J Neurol Neurosurg Psychiatry 49: 657–660

Hakim AM, Mathieson G (1978) Basis of dementia in Parkinson's disease. Lancet ii: 729

Hakim AM, Mathieson G (1979) Dementia in Parkinson disease: a neuropathologic study. Neurology 29: 1209–1214

Halgin R, Riklan M, Misiak H (1977) Levodopa, Parkinsonism, and recent memory. J Nerv Ment Dis 164: 268–272

Hare M (1978) Clinical check list for diagnosis of dementia. Br Med J 2: 266–267

Hart RP, Kwentus JA (1987) Psychomotor slowing and subcortical-type dysfunction in depression. J Neurol Neurosurg Psychiatry 50: 1263–1266

Haxby JV, Grady ChL, Duara R, et al (1986) Neocortical metabolic abnormalities precede nonmemory cognitive defects in early Alzheimer's type dementia. Arch Neurol 43: 882–885

Heilig ChW, Knopman DS, Mastri AR, Frey W (1985) Dementia without Alzheimer pathology. Neurology 35: 762–765

Heston LL (1980) Dementia associated with Parkinson's disease: a genetic study. J Neurol Neurosurg Psychiatry 43: 846–848

Hietanen M, Teräväinen H (1984) Psychomotor and cognitive performance in Parkinsonism. Acta Neurol Scand 69: 61–62

Hoff P, Hippius H (1989) Alois Alzheimer 1864–1915. Nervenarzt 60: 332–337

Huber SJ, Shuttleworth EC, Paulson GW, et al (1986) Cortical vs subcortical dementia/ Neuropsychological differences. Arch Neurol 43: 392–394

Huber SJ, Shulman HG, Paulson GW, Shuttleworth EC (1989) Dose-dependent memory impairment in Parkinson's disease. Neurology 39: 438–440

Javoy-Agid F, Agid Y (1980) Is the mesocortical dopaminergic system involved in Parkinson disease? Neurology 30: 1326–1330

Kish SJ, Chang LJ, Mirchandani L, et al (1985) Progressive supranuclear palsy: relationship between extrapyramidal disturbances, and brain neurotransmitter markers. Ann Neurol 18: 530–536

Lauter H, Kurz A, Zimmer R (1986) Klinische Probleme bei der Diagnostik von Demenzprozessen im Alter. Act Neurol 13: 11–18

Lees AJ (1985) Parkinson's disease and dementia. Lancet i: 43–44

Lees AJ (1989) Neuropsychologische Störungen beim Morbus Parkinson. Nervenarzt 60: 71–79

Lees AJ, Smith E (1983) Cognitive deficits in the early stages of Parkinson's disease. Brain 106: 257–270

Leverenz J, Sumi SM (1986) Parkinson's disease in patients with Alzheimer's disease. Arch Neurol 43: 662–664

Levin BE, Llabre MM, Weiner WJ (1989) Cognitive impairments associated with early Parkinson's disease. Neurology 39: 557–561

Lewin R (1985) Parkinson's disease: an environmental cause? Science 229: 257–258

Liebermann A, Dziatolowski M, Kupersmith M, et al (1979) Dementia in Parkinson's disease. Ann Neurol 6: 355–359

Loranger AW, Goodell H, McDowell FH (1972) Intellectual impairment in Parkinson's disease. Brain 95: 405–412

Mann JJ, Stanley M, Kaplan RD, et al (1983) Central catecholamine metabolism in vivo and in cognitive and motor deficits in Parkinson's disease. J Neurol Neurosurg Psychiatry 46: 905–910

Martin WE, Loewenson RB, Resch JA, Baker AB (1973) Parkinson's disease. Clinical analysis of 100 patients. Neurology 23: 783–790

Matthews ChG, Haaland KY (1979) The effect of symptom duration on cognitive and motor performance in parkinsonism. Neurology 29: 951–956

Mayeux R, Stern Y, Rosen J, Leventhal J (1981) Depression, intellectual impairment, and Parkinson disease. Neurology 31: 645–650

Mayeux R, Stern Y, Spanton S (1985) Heterogeneity in dementia of the Alzheimer type: evidence of subgroups. Neurology 35: 453–461

Mayeux R, Stern Y, Rosen J, Benson DF (1983) Is "Subcortical dementia" a recognizable clinical entity? Ann Neurol 14: 278–283

Mayeux R, Stern Y, Rosen J, Benson DF (1981) Subcortical dementia: a recognizable clinical entity. Ann Neurol 10: 100–101

Mindham RHS, Ahmed SAW, Clongh C (1982) A controlled study of dementia in Parkinson's disease. J Neurol Neurosurg Psychiatry 45: 969–974

Mindham RHS (1970) Psychiatric symptoms in Parkinsonism. J Neurol Neurosurg Psychiatry 33: 188–191

Mohr E, Fabbrini G, Williams J, et al (1989) Dopamine and memory function in Parkinson's disease. Movement Dis 4: 113–120

Monte SM, Wells SE, Hedley-Whyte ET, Growdon JH (1989) Neuropathological distinction between Parkinson's dementia and Parkinson's plus Alzheimer's disease. Ann Neurol 26: 309–320

Mortimer JA, Pirozzolo FJ, Hausch EC, Webster DD (1982) Relationship of motor symptoms to intellectual deficits in Parkinson disease. Neurology 32: 133–137

Mortimer JA, Hausch EC, Pirozzolo FJ, Webster DD (1982) Continuum of intellectual deficit in Parkinson disease. Ann Neurol 12: 402–403

Mortimer JA, Christensen KJ, Webster DD (1985) Parkinsonian dementia. Handbook Clin Neurol 46: 371–384

Oyebode JR, Barker WA, Blessed J, et al (1986) Cognitive functioning in Parkinson's disease in relationship to prevalence of dementia and psychiatric diagnose. Br J Psychiatry 149: 720–725

Pearce J (1974) The extrapyramidal disorder of Alzheimer's disease. Eur Neurol 12: 94–103

Perry RH, Tomlinson BE, Candy JM, et al (1983) Cortical cholinergic deficit in mentally impaired parkinsonian patients. Lancet ii: 789–790

Perry EK, Curtis M, Dick DJ, et al (1985) Cholinergic correlates of cognitive impairment in Parkinson's disease: comparison with Alzheimer's disease. J Neurol Neurosurg Psychiatry 48: 413–421

Phelps EM, Mazziotta JC, Schelbert HR (1986) PET and autoradiography. Principles and applications for the brain and heart. Raven Press, New York

Piccirilli M, Piccinin Gl, Agostini L (1984) Characteristic clinical aspects of parkinson patients with intellectual impairment. Eur Neurol 23: 44–50

Pillon B, Dubois B, Lhermitte F, Agid Y (1986) Heterogeneity of cognitive impairment in progressive supranuclear palsy. Neurology 36: 1179–1185

Pillon B, Dubois B, Cusimano G, et al (1989) Does cognitive impairment in Parkinson's disease result from non-dopaminergic lesion. J Neurol Neurosurg Psychiatry 52: 201–206

Pollock M, Hornabrook RW (1966) The prevalence, natural history and dementia of Parkinson's disease. Brain 89: 429–448

Portin R, Rinne UK (1984) Predictive factors for dementia in Parkinson's disease. Acta Neurol Scand 69: 57–58

Rafal RD, Posner MJ, Walker JA, Friedrich FJ (1984) Cognition and the basal ganglia. Brain 107: 1083–1094

Rapcsak SZ, Croswell SC, Rubens AB (1989) Apraxia in Alzheimer's disease. Neurology 39: 664–668

Rinne UK, Laakso K, Mölsä P, et al (1984) Relationship between Parkinson's and Alzheimer's disease. Involvement of extrapyramidal, dopaminergic, cholinergic and somatostatin mechanism in relation to dementia. Acta Neurol Scand 69: 59–60

Rogers D, Lees AJ, Smith E, et al (1987) Bradyphrenia in Parkinson's disease and psychomotor retardation in depressive illness. Brain 110: 761–776

Rosenberg RN, Green JB, White ChL, et al (1989) Dominantly inherited dementia and parkinsonism, with non-Alzheimer amyloid plaques: a new neurogenetic disorder. Ann Neurol 25: 152–158

Sala S, Lorenzo G, Giordano A, Spinnler H (1986) Is there a specific visuo-spatial impairment in Parkinsonians? J Neurol Neurosurg Psychiatry 49: 1258–1265

Scatton B, Rouquier L, Javoy-Agid F, Agid Y (1982) Dopamine deficiency in the cerebral cortex in Parkinson's disease. Neurology 32: 1039–1040

Shankar SK, Yanagihara R, Garruto RM, et al (1989) Immuncytochemical characterisation of neurofibrillary tangles in amyotrophic lateral sclerosis and parkinsonism-dementia of Guam. Ann Neurol 25: 146–151

Smet Y, Ruberg M, Serdaru M, et al (1982) Confusion, dementia, and anticholinergics in Parkinson's disease. J Neurol Neurosurg Psychiatry 45: 1161–1164

Sroka H, Elizan MD, Yahr MD, et al (1981) Organic mental syndrome and confusional states in Parkinson's disease. Arch Neurol 38: 339–342

Starkstein S, Leignarda R, Gershanik O, Berthier M (1987) Neuropsychological disturbances in hemiparkinson's disease. Neurology 37: 1762–1764

Stern Y, Langston JW (1985) Intellectual changes in patients with MPTP-induced parkinsonism. Neurology 35: 1506–1509

Summers WK, Viesselman JO (1981) Use of THA in treatment of Alzheimer-like dementia: pilot study in 12 patients. Biol Psychiatry 16: 145–153

Sweet RD, McDowell FH, Feigenson JS, et al (1976) Mental symptoms in Parkinson's disease during chronic treatment with levodopa. Neurology 26: 305–310

Talland GA (1962) Cognitive function in Parkinson's disease. J Nerv Ment Dis 135: 196–205

Taylor AE, Saint-cyr JA, Lang AE (1986) Frontal lobe dysfunction in Parkinson's disease. Brain 109: 845–883

Torack RM, Morris JC (1986) Mesolimbocortical dementia. A clinicopathologic case study of a putative disorder. Arch Neurol 43: 1074–1078

Villardita C, Smirni P, Pira F, et al (1982) Mental deterioration, visuoperceptive disabilities and constructional apraxia in Parkinson's disease. Acta Neurol Scand 66: 112–120

Wetterling T (1989) Alzheimersche Krankheit: Überblick über den aktuellen Stand der Forschung. Fortschr Neurol Psychiat 57: 1–13

Whitehouse PJ (1986) The concept of subcortical and cortical dementia: another look. Ann Neurol 19: 1–6

Whitehouse PJ (1986) From: Alzheimer's and Parkinson's diseases. Plenum Publishing Corporation, pp 85–94

Whitehouse PJ (1989) Parkinson's disease and Alzheimer's disease: new neurochemical parallels. Movement Dis 4: 57–62

Whitehouse PJ, Unnerstall JR (1988) Neurochemistry of dementia. Eur Neurol 28: 36–41

Whitehouse PJ, Price DL, Clark AW, et al (1981) Alzheimer's disease: evidence for selective loss of cholinergic neurons in the nucleus basalis. Ann Neurol 10: 122–126

Whitehouse PJ, Hedreen JC, White ChL, Price DL (1983) Basal forebrain neurons in the dementia of Parkinson's disease. Ann Neurol 13: 243–248

Whitehouse PJ, Vale WW, Zweig RM, et al (1987) Reduction in corticotropin releasing factor-like immunoreactivity in cerebral cortex in Alzheimer's disease, Parkinson's disease, and progressive supranuclear palsy. Neurology 37: 905–909

Whitehouse PJ, Martino AM, Wagster MV, et al (1988) Reduction in [3H] nicotinic acetylcholine bindings in Alzheimer's disease and Parkinson's disease: an autoradiographic study. Neurology 38: 720–723

Yoshimura M (1983) Cortical changes in the parkinsonian brain: a contribution to the delineation of "diffuse Lewy body disease". J Neurol 229: 17–32

Zweig RM, Shegg KM, Peacock JH, Melarkey D (1989) A case of Alzheimer's disease and hippocampal sclerosis with normal cholinergic activity in the basal forebrain, neocortex, and hippocampus. Neurology 39: 288–290

Authors' address: I. Marschall, M.D., Neurological University Clinic, Lazarettgasse 14, A-1090 Vienna, Austria

J Neural Transm (1991) [Suppl] 33: 93–97
© by Springer-Verlag 1991

Memory dysfunction in Parkinson patients: an analysis of verbal learning processes

E. Karamat, J. Ilmberger, W. Poewe, and **F. Gerstenbrand**

Neurological Clinic, University of Innsbruck, Austria

Summary. The German version of the California verbal learning test (Münchner Gedächtnis-Test) was administered to compare memory functions of 45 Parkinson patients and 32 healthy controls. Although the Parkinson patients were within their age norm in their performance on standard intelligence and memory tests, a significant pattern of impaired learning was observed with the MGT.

While learning capacity showed a linear increase, it was altogether lower than that of healthy normals.

Parkinson patients have difficulty organizing new material and applying useful strategies.

When offered useful cues, they can apply these for a while, but soon loose these strategies again. They are more irritated by distractors than healthy people. Once material is learned and stored, it hampers the learning of new material. Also, new information disturbs the reproduction of formerly acquired information.

The ability to learn and to store new material demands a variety of cognitive processes, which again are associated with different brain regions. Specialized neuropsychological tests have recently shown that different patterns of memory impairment can be distinguished. Great interest is therefore focused on finding an instrument with which to observe and quantify some of the cognitive processes necessary for learning, storing and reproducing information.

Test procedure

An experimental German memory test based on the structure of the California verbal learning test was used. The test basically consisted of three lists: list A, list B and a recognition list. List A was composed of 16 shopping items from four semantic categories. List B had the same structure; two semantic categories were the same as in list A. The recognition list was composed of the items on lists A and B and distractor items.

List A was presented to each test subject five times; the task demanded that as many items as possible be named from memory in any given order. Responses were recorded on each trial. After that, list B was presented with the same instructions. Then the subject was asked to name as many items as possible from list A. Following this the subject was confronted with the category names from list A and again asked to name as many items as he could remember. After a break of 20 minutes, the subject was once more asked for the items on list A, with the category names of list A again being given as cues. The recognition list was also presented.

Among other components, the following can be evaluated: correct answers, semantic clustering score (one point for each item following another of the same semantic category), amount of learning in trials 1 to 5, proactive and retroactive interference, effects of delay on recall and effects of semantic cueing.

Subjects

The collective consisted of 45 idiopathic Parkinson patients (17 female, 28 male) with a mean age of 59.7 years. Mean duration of L dopa therapy was 3.7 years and mean L dopa dose 524 mg/day. Mean H + Y was 2. Mean total motor score was 13 ("on") (UPRS). CT scanning showed 28 to be normal and seven to have a slight atrophy. Mean WAIS IQ was 102, VIQ 105 and PIQ 98. The Wechsler memory scale showed a logic memory of 94.8, a digit span of 99.7 and an associate learning of 99.5.

Performance of the Parkinson patients in the above-mentioned standard tests was well within their age norm. The control group consisted of 32 healthy persons with an average age of 52.2 years. No neuropsychological tests were administered.

Results

ANOVA's for repeated measures were calculated for the two experimental groups and, with a grouping factor, for all subjects. Subsequently, the significance of differences between adjacent trials was tested.

Mean correct answer scores were obtained for both experimental groups for trials 1 to 5. In general the Parkinson group performed significantly worse than the control group ($p < .0005$). For the control group each increase is significant, reading in a linear trend; for the Parkinson group, only the increases up to the fourth trial are significant (Fig. 1).

The mean semantic clustering scores for trials 1 to 5 showed significantly lower semantic clustering for Parkinson patients than for the control group ($p < .0005$). In the control group, semantic clustering increased significantly up to the fourth trial; for the Parkinson group, the increases were significant only up to the third trial (Fig. 2).

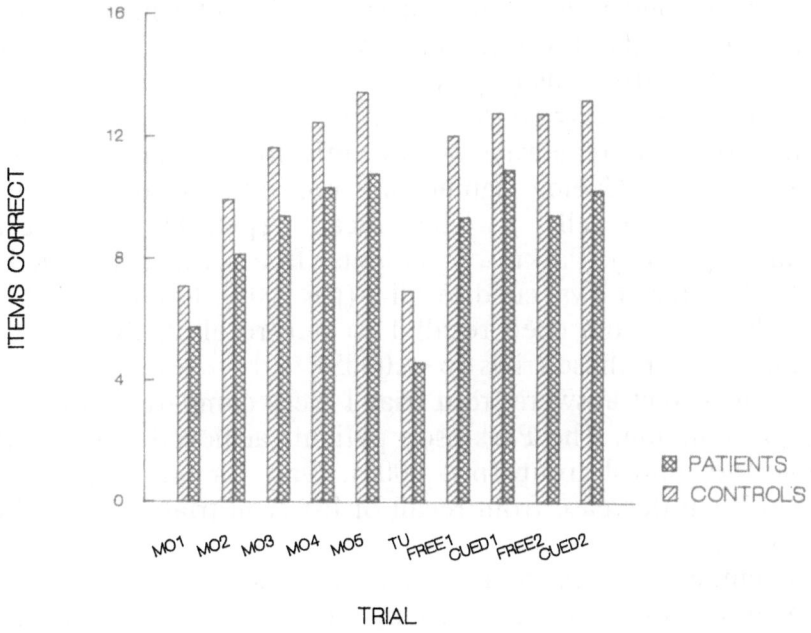

Fig. 1

The mean correct answers on trial 5 were compared with those of the subsequent free recall trials. Parkinson patients always performed significantly worse than the control group (p < .0005). For the control group, scores between trial 5 and the first free recall trial did not differ significantly, but the

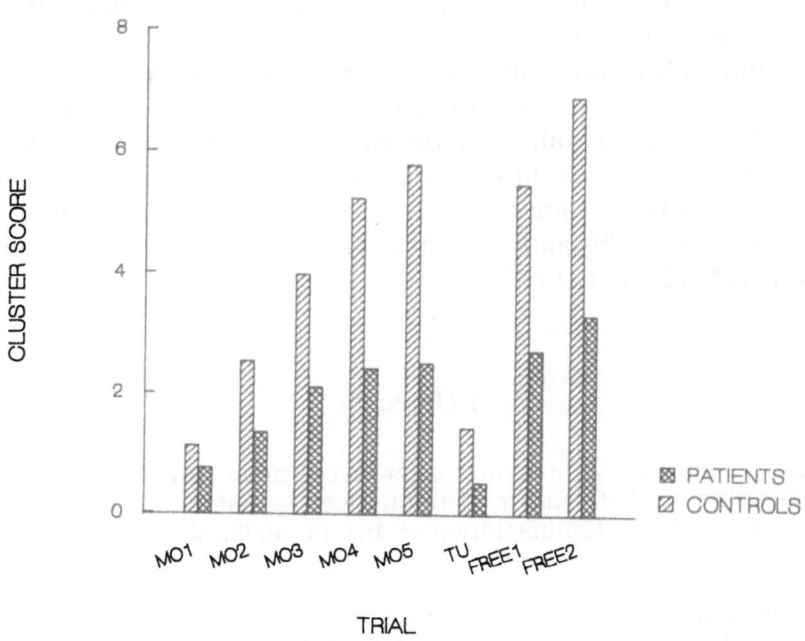

Fig. 2

scores for the second free recall trial are significantly higher than for the preceding trials (p < .005). For the Parkinson group, there was no significant difference between the trials (Fig. 1).

The mean correct answers for the free versus the cued recall trials in Parkinson patients were always significantly lower than those in control subjects (p < .0005). Cueing significantly improved recall for both groups (control group; free recall 1 vs. cued recall 1 (p < .005), free recall 2 vs. cued recall 2 (p < .05); Parkinson patients: free recall 1 vs. cued recall 1 (p < .0005), free recall 2 vs. cued recall 2 (p < .005). In addition there was a significant decrease from cued recall 1 to free recall 2, the twenty-minute break being between those trials (p < .0005).

The mean correct answers from trial 1 were compared with those from the list 2 presentation. The Parkinson patients achieved significantly lower scores than the control group (p < .0005). Only for the patient group was there a significant decrease from recall of list A in trial 1 to recall of list B (p < .005) (Fig. 1).

The semantic clustering scores for trial 5 and the two free recall trials were compared. For the control group, there was a significant increase in semantic clustering from the first to the second free recall trial (p < .005), whereas the scores of the Parkinson patients did not differ significantly between trial 5 and free recall trial 1 and between free recall trial 1 and free recall trial 2 (Fig. 2).

Discussion

The so often reported memory impairment in Parkinson patients might be more a learning impairment. Lack of strategies, inability to organize new material could well result from a lower level of processing, or a decreased speed of information processing. Proactive and retroactive interference could result from "shifting" inability, or the inhability to select relevant information and to suppress irrelevant information. These impairment patterns have been described as typical impairments found in the normal old-age population (Fleischmann, 1989) and were also observed in a population of frontal lobe lesions (Hambrecht, 1987).

References

Fleischmann U (1989) Gedächtnis und Alter. Multivariate Analysen zum Gedächtnis alter Menschen, 1. Aufl. Huber, Bern Stuttgart Toronto
Hambrecht M (1987) Gedächtnisstörungen bei Frontalhirnläsionen. Nervenarzt 58: 131–136

not cited in the text:
v Cramon D, Zihl J (eds) (1988) Neuropsychologische Rehabilitation. Springer, Berlin Heidelberg New York

Delis DC, Kramer J, Ober AB (1986) The California verbal learning test: administration and interpretation. Boston V.A. Medical Center
Delis DC, Butters N, et al (1988) Wechsler memory scale revised and CVLT: convergence and divergence. Clin Neuropsychol 2: 188–196
Meier M, Benton AL, Diller L (eds) (1987) Neuropsychological rehabilitation. Churchill Livingstone, New York
Pope DM (1987) The California VLT: performance of normal adults aged 55–91. California School of Professional Psychology, Berkeley

Authors' address: Dr. E. Karamat, Ph.D., University Neurological Clinic, Anichstrasse 35, A-6020 Innsbruck, Austria

J Neural Transm (1991) [Suppl] 33: 99–103
© by Springer-Verlag 1991

Does the absence of clinical expression of choreoathetosis, despite severe striatal atrophy, correlate with plasticity of neuropeptide synthesis?

S. N. Schiffmann and **J.-J. Vanderhaeghen**

Laboratory of Neuropathology and Neuropeptide Research, Faculty of Medicine, Erasme and Brugmann Hospitals, Université Libre de Bruxelles, Brussels, Belgium

Summary.   Neuropeptide and neurotransmitter plasticity has been demonstrated in the central nervous system. Modifications of their synthesis occur following receptor blockade or deafferentiation by surgical lesions. This concept should provid answers to some remaining open questions in human pathology especially in degenerative diseases of the basal ganglia. In a severely atrophied striatum we observed a selective increase in the number of detectable striatal substance P and met-enkephalin neurones which exhibited a striking increase in the intensity of labelling. This increase, instead of the well established reduction of substance P and enkephalins in the atrophied striatum of Huntington's disease, could explain the absence of choreoathetosis which was replaced by rigidity and bradykinesia in the patient. The absence of choreoathetosis, despite severe striatal atrophy, is described in several basal ganglia diseases and could also be related to neurotransmitter or neuropeptide plasticity rather than due to the primary lesion.

Modifications of neurotransmitter or neuropeptide expression by receptor blockade or lesions of deafferentiation have been described in the central nervous system (Bannon et al., 1986; Normand et al., 1988). This plasticity exists in numerous systems and has been extensively demonstrated in the basal ganglia. Indeed, the expression of several striatal neuropeptides such as met-enkephalin (Met-Enk) (Normand et al., 1988), substance P (SP) (Bannon et al., 1986), neuropeptide Y (NPY) or dynorphin is modulated, differentially, by the nigrostriatal dopaminergic pathway probably at the transcriptional level.

Whilst the primary neurochemical defect in some degenerative diseases of the basal ganglia is well established; dopamine in Parkinson's disease or gamma-aminobutyric acid (GABA) in Huntington's disease (HD) (Bird and Iversen, 1974); it could not be excluded that part of the symptomatology was related to secondary modifications of neurotransmitter or neuropeptide expression in neuropathologically intact target structures.

HD is dominantly inherited and classically characterized by chorea associated with dementia (Bruyn and Went, 1986). A severe striatal atrophy and a 50–80% neuronal loss in the caudate-putamen constitute its main neuropathological features (Roos, 1986). Several other neurological diseases (Greenfield and Wolfsohn, 1922; Wilson, 1912) are also characterized by lesions of the striatum associated with abnormal movements such as choreoathetosis. Neurochemical studies in HD have previously established a reduction of specific markers of striatal medium-sized spiny neurones such as GABA (Bird and Iversen, 1974; Spokes, 1980), Met-Enk (Emson et al., 1980; Ferrante et al., 1987b) and (SP) (Emson et al., 1980; Ferrante et al., 1987b) which resulted in a GABA (Bird and Iversen, 1974; Spokes, 1980), SP (Emson et al., 1980) and Met-Enk (Emson et al., 1980) depletion in the globus pallidus and the substantia nigra pars reticulata.

In some rare cases of juvenile and adult HD, the Westphal variant (Bird and Pauslon, 1971; Bruyn and Went, 1986), as well as in other striatal lesions (Adams et al., 1961; Peters et al., 1979), choreoathetosis is transient or absent without any apparent explanation. Links between such peculiar clinical entities and the new concept of neurotransmitter-neuropeptide plasticity have yet to be established but could be relevant in understanding them. In that perspective we have decribed (Schiffmann and Vanderhaeghen, 1989) the case of a 49-year old patient who presented a progressive intellectual deterioration evolving into severe dementia until his death five years later. Two years before his death he developed bradykinesia and rigidity unresponsive to levodopa, whilst tremor, neither choreic nor athetotic movements was not observed. One year before his death he developed a generalized hyperreflexia and a left upper limb paresis. The family history indicates the following: the mother is a healthy 77-year old woman. The father who had no known record of familial neuropsychiatric diseases, was alcoholic and had committed suicide at the age of 76. Of the 4 siblings, 3 were alive and healthy, the fourth was the propositus. In addition, late in their lives 2 sisters of the mothers developed severe dementia.

In this case, the caudate nucleus and the putamen were severely shrunken (Fig. 1B) and exhibited a dense fibrillary gliosis in addition to the neuronal loss. Neuropathological lesions were not found elsewhere in the brain not even in the cerebral cortex and the substantia nigra.

As studied by immunocytochemistry, there was macroscopically a significant increase of SP and Met-Enk staining in the atrophied striatum (Fig. 1B). The global increase of staining was due not only to greater intensity of labelling of the individual neurones but also to an increase in the number per field of detectable nerve cell bodies and fibres containing these peptides (Fig. 1C). The compartmental striosome/matrix organization of these peptides is well preserved (Fig. 1A, B). Simultaneously, no changes were observed in the dense meshworks of SP and enkephalin fibres, found respectively in the medial and lateral globus pallidus and in the substantia nigra pars reticulata. On the contrary, for acetyl cholinesterase (AChE), somatostatin (SS) and NPY present essentially in the matrix, the intensity of staining appeared macroscopically unchanged. Neurotensin (NT) and cholecystokinin (CCK) thin nerve fibres were present with a normal appearance and distribution.

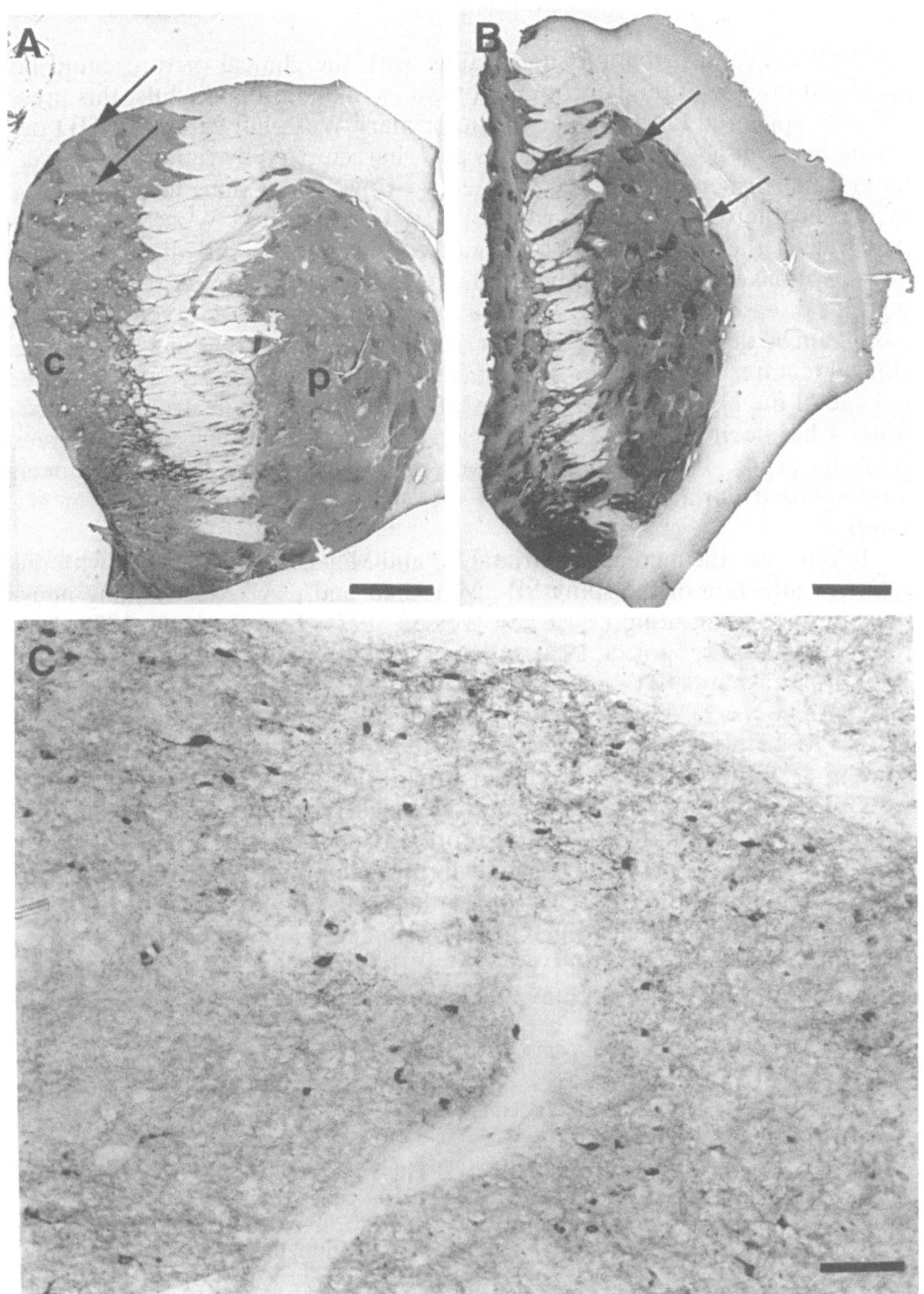

Fig. 1. Sections from normal (**A**) and diseased (**B**) striata taken at the level of the caudato-putaminal junction immunolabelled for Met-Enk. (**B**) A major atrophy of the caudate and putamen is associated with an increase of Met-Enk staining. The compartmental organisation of Met-Enk in striosomes (arrows) is similar in the normal and atrophied striata. (**C**) High densities of intensely labeled Met-Enk containing neurons are present in the anterior and dorsal diseased putamen. Abbreviations: *c* caudate nucleus; *p* putamen. Bars = 5 mm in A, B and 100 um in C

Whilst several conditions compatible with the clinical picture could be ruled out in view of the neuropathological examination and whilst this present case appeared fairly similar to the primary Westphal variant of HD occuring in about 5% of the adult cases and characterized by rigidity, akinesia, mental impairment and a high frequency of pyramidal signs in the absence of choreoathetosis (Bird and Paulson, 1971; Bruyn and Went, 1986), the pedigree of the present family was not consistent with this diagnosis. The lacking of this latter critical aspect led us to consider a new mutation to Huntington's disease (Wolff et al., 1989) or an yet undescribed or atypical form of basal ganglia degeneration. However, it should be noted that in a family with clinical features similar to the Westphal variant and an onset in adult life, the linkage of the DNA G8 probe to the HD locus on the short arm of chromosome 4 has been reported (Zweig et al., 1989) as in classical HD pedigrees (Gusella et al., 1983). Moreover, most members of this family had been misdiagnosed during life because of the absence of chorea (Zweig et al., 1989).

In our case, the increase of striatal SP and Met-Enk is in contrast with the selective affection of the spiny SP-, Met-Enk- and GABA-containing neurones in the HD striatum (Bird and Iversen, 1974; Emson et al., 1980; Ferrante et al., 1987b; Spokes, 1980) where the AChE neurones (Ferrante et al., 1987a) and aspiny NPY- and SS-containing neurones (Aronin et al., 1983; Dawbarn et al., 1985) were spared. Such results could reflect an increase of SP and Met-Enk synthesis or a defect in their release and therefore a higher content of these peptides per cell in some selectively spared striatal neurones. This increase of synthesis could be itself secondary to a peculiar undetermined lesion and could then represent an example of neuropeptide plasticity in human pathology. Clinically, it is tempting to attribute the absence of chorea to this increase of synthesis of SP and Met-Enk in some preserved striatal neurones since in HD with expression of chorea, a reduction of these peptides is observed. The differences in neuropeptide and/or neurotransmitter expression may therefore explain different clinical features of degenerative diseases.

References

Adams RD, Van Bogaert L, Vendereecken H (1961) Dégénérescences nigro-striées et cérebellonigro-striées. Psychiat Neurol 142: 219–259

Aronin N, Cooper PE, Lorenz LJ, Bird ED, Sagar SM, Leeman SE, Martin JB (1983) Somatostatin is increased in the basal ganglia in Huntington's disease. Ann Neurol 13: 519–526

Bannon MJ, Lee J-M, Giraud P, Young A, Affloter H-U, Bonner TI (1986) Dopamine antagonist haloperidol decreases substance P, substance K and preprotachykinin mRNAs in rat striatonigral neurons. J Biol Chem 261: 6640–6642

Bird ED, Iversen LL (1974) Hungtington's chorea: post-mortem measurement of glutamic acid decarboxylase, choline acetyltransferase and dopamine in basal ganglia. Brain 97: 457–472

Bird MT, Pauslon GW (1971) The rigid form of Huntington's chorea. Neurology 21: 271–276

Bruyn GW, Went LN (1986) Huntington's chorea. In: Vincken PJ, Bruyn GW, Klawans HL (eds) Handbook of clinical neurology, vol 5. Extra-pyramidal disorders. Elsevier, Amsterdam, pp 267–313

Dawbarn D, De Quidt ME, Emson PC (1985) Survival of basal ganglia neuropeptide Y-somatostain neurones in Huntington's disease. Brain Res 340: 251–260

Emson PC, Arregui A, Clement-Jones V, Sandberg BEB, Rossor M (1980) Regional distribution of methionine-enkephalin and substance P-like immunoreactivity in normal human brain and in Huntington's disease. Brain Res 199: 147–160

Ferrante RJ, Beal MF, Kowall NW, Richardson EP Jr, Martin JB (1987a) Sparing of acetylcholinesterase-containing neurons in Huntington's disease. Brain Res 411: 162–166

Ferrante RJ, Kowall NW, Richardson EP Jr, Bird ED, Martin JB (1987b) Topography of enkephalin, substance P and acetylcholinesterase staining in Huntington's disease striatum. Neurosci Lett 71: 283–288

Greenfield J, Wolfsohn J (1922) The pathology of Syndenham's chorea. Lancet ii: 603–606

Gusella JF, Wexler NS, Conneally PM, Naylor SL, Anderson MA, Tanzi RE, Watkins PC, Ottina K, Wallace MR, Sakaguchi AY, Young AB, Shoulson I, Bonnilla E, Martin JB (1983) A polymorphic DNA marker genetically linked to Huntington's disease. Nature 306: 234–238

Normand E, Popovici T, Onteniente B, Fellman D, Piatier-Tonneau D, Auffray C, Bloch B (1988) Dopaminergic neurons of the substantia nigra modulate preproenkephalin A gene expression in rat striatal neurons. Brain Res 439: 39–46

Peters ACB, Vielvoye GJ, Versteeg J, Bots GT, Lindeman J (1979) ECHO 25 focal encephalitis and subacute hemichorea. Neurology 29: 678–681

Roos RAC (1986) Neuropathology of Huntington's chorea. In: Vincken PJ, Bruyn GW, Klawans HL (eds) Handbook of clinical neurology, vol 5. Extrapyramidal disorders. Elsevier, Amsterdam, pp 315–326

Schiffmann SN, Vanderhaeghen J-J (1989) Increase of substance P and met-enkephalin in a severely atrophied striatum without clinical expression of chorea. Neurochem Int 14: 175–183

Spokes EGS (1980) Neurochemical alterations in Huntington's chorea. A study of postmortem brain tissue. Brain 103: 179–210

Wilson SAK (1912) Progressive lenticular degeneration: a familial nervous system disease associated with cirrhosis of the liver. Brain 34: 295–507

Wolff G, Deuschl G, Wienker TF, Hummel K, Bender K, Lucking CH, Schumacher M, Hammer J, Oepen G (1989) New mutation to Huntington's disease. J Med Genet 26: 18–27

Zweig RM, Koven SJ, Hedreen JC, Maestri NE, Kazazian Jr HH, Folstein SE (1989) Linkage to the Huntington's disease locus in a fanily with unusual clinical and pathological features. Ann Neurol 26: 78–84

Authors' address: Dr. S.N. Schiffmann, Laboratory of Neuropathology and Neuropeptide Research, Université Libre de Bruxelles, Campus Erasme, CP 601, 808 route de Lennik, 1070 Brussels, Belgium.

J Neural Transm (1991) [Suppl] 33: 105–110
© by Springer-Verlag 1991

Essential tremor: functional disability vs. subjective impairment

E. Auff, A. Doppelbauer, and **E. Fertl**

Neurological Clinic, University of Vienna, Austria

Summary.   78 patients with essential tremor (ET) were investigated to un-cover correlation and discrepancies between functional (motor) disabilities and subjective impairment. Various self-rating scales (Zung, v. Zerssen etc.) were used for the assessment of the latter: 2/5 of the patients rated them-selves as severely impaired; 1/3 was depressive. Patients who showed nearly the same functional (motor) disability felt very differently subjectively im-paired. Semiquantitative clinical scores of action tremor correlated best with the subjective impairment in activities of daily living. Objective measure-ments of motor disability were performed with the "Motorische Leistungs-serie nach Schoppe" (motor performance test) and showed good correlation to the subjective impairment in simple tasks of every day life, such as drink-ing from a glass, eating soup, and writing. Asking for the subjective impair-ment in these tasks allows to estimate the objective disability correctly. This may be of value in long-term studies of essential tremor.

Introduction

Essential tremor (ET) is generally known as the most common movement disorder (Rautakorpi et al., 1984). It is characterized by a postural tremor of the hands, sporadic or familiar occurence, frequent improvement after alcohol ingestion and response to treatment with beta-receptor blocking agents and/or primidone (Marsden, 1984). Only a rather small number of patients with ET feels impaired or disabled by the disease (Sutherland et al., 1975), approximately 10 percent of the patients do seek medical help (Rau-takorpi et al., 1982). The severity of the tremor shows considerable interin-dividual, but also intraindividual fluctuations (Koller and Royse, 1985). The subjective impairment due to the tremor, however, is very differently per-ceived by the patients.

ET has to be differentiated from various other forms of tremor. Patients suffering from ET are often misdiagnosed as having Parkinson's disease, perhaps because Parkinson patients may also show an additional or even sole postural tremor beside their classical resting tremor.

The purpose of our study was to try to answer the following questions:
— Which factors may contribute to the well-known inter- and intraindividual fluctuations of tremor intensity?
— Which of the subjective impressions of the patient about his impairment due to the tremor are correlating best to objective measurements of motor disability?
— Are there any neurological symptoms found in patients with essential tremor, that may lead to confusion with Parkinson's syndrome?

Patients and methods

78 patients with "pure" essential tremor were investigated (age 68.8 ± 12.8 years; mean ± SD; 32 males, 68.6 ± 13.7 years; 46 females, 68.9 ± 12.4 years).

All patients were investigated in the following way: clinical history, complete neurological investigation using a semiquantitative scale (similar to and including all items of the Columbia University Rating Scale for Parkinson's disease); psychiatric exploration (with special respect to depressive disorders and/or organic brain syndrome); "Motorische Leistungsserie nach Schoppe" for tremor quantification and as a motor performance test; psychosocial status; neuropsychological investigations (including various self-rating scales, light flickering frequency analysis, Benton-test, and Rorschach-test).

As a tool for objective measurement the "Motorische Leistungsserie nach Schoppe (MLS)" and the semiquantitative clinical neurological investigation were used. Out of the whole motor performance test (MLS) the four most relevant subtests for tremor patients were chosen: steadiness, line following, tapping, aiming. From the clinical neurological investigation the items head tremor; resting tremor, postural tremor, intention tremor (all from the upper limbs) were used for comparison.

To register subjective impairment various self-rating scales (Self-rating Depression Scale Zung, Beschwerdenliste v. Zerssen, Befindlichkeitsskala v. Zerssen, visual analogue scales for subjective impairment of various activities of daily living) were used.

Results

Results are summarized in Tables 1–3.

Patients with ET did not only present a postural tremor, but often an additional resting and/or intention tremor component (Table 1). Patients (also) having head tremor often showed a scanning speech. Among the older patients mild rigidity and/or another motor problem, such as mild bradykinesia was found, without fulfilling all the criteria for Parkinson's disease.

1/3 of the patients had a noticeable disturbance of dexterity or motor performance, which could not be ascribed to the tremor alone.

Comparing the results of the (objective) measuring methods and the results of the patients' subjective estimation (Tables 2 and 3) we found no conformity between tremor intensity and various well-known self-rating scales (SDS-Zung, BF-v. Zerssen, B-L-v. Zerssen).

In the clinical investigation the intensity of postural tremor correlates best to the subjectively found disability in every-day life tasks, other aspects of motor disturbance, like resting tremor or intention tremor may contribute

Table 1. Semiquantitative neurological investigation (modified Columbia University Rating Scale for Parkinson's disease). Scores: 0 = absent . . . 4 = severe

	n	0	Scores 1/2	3/4
Head tremor	78	33	36	9
Resting tremor				
right hand	78	48	26	4
left hand	78	45	31	2
right leg	78	70	7	1
left leg	78	73	4	1
Postural tremor				
right hand	78	4	56	18
left hand	78	4	63	11
right leg	78	65	12	1
left leg	78	67	10	1
Intention tremor				
right arm	78	36	40	2
left arm	78	36	41	1
Rigidity				
Neck	78	70	8	0
right arm	78	73	5	0
left arm	78	74	4	0
right leg	78	74	4	0
left leg	78	75	3	0
Finger dexterity				
right	77	53	24	0
left	78	57	21	0
Pronation/supination				
right hand	77	66	11	0
left hand	78	65	13	0
Speech	78	58	17	3
Posture	78	61	15	2
Gait abnormality	78	63	14	1
Bradykinesia	78	56	20	2

to the impairment (Table 2). Head tremor does not play any role in motor disturbance, but may be generally troublesome.

On the other hand there was a very good correspondence between objective tremor measurements and the patients' statements about their subjective impairment (semiquantitative assessment). The results of the subtests of the MLS showed good correlation with subjectively perceived motor disturbance of simple, clearly defined tasks of every-day life (writing a letter, signature,

Table 2. Correlation of results from the clinical score (semiquantitative neurological investigation; *rest* resting tremor, *post* postural tremor, *int* intention tremor) and various self-rating scales (*BFS* Befindlichkeitsskala v. Zerssen, *BL* Beschwerdenliste v. Zerssen, *SDS* Self-rating Depression Scale Zung, *VAS* visual analogue scale)

| | Tremor | | | |
	head	rest	post	int
BFS v. Zerssen	−	−	−	−
BL v. Zerssen	−	−	−	−
SDS-Zung	−	−	−	−
VAS eating	−	−	+	+
VAS soup	−	−	+	−
VAS drinking	−	−	+	−
VAS writing	−	−	+	+
VAS signature	−	−	+	+
VAS dressing	−	+	+	+
VAS hygiene	−	+	+	−
VAS working	−	−	−	−
VAS hobbies	−	−	−	−

+ statistically significant correlation ($p < 0.05$), − n.s.

eating, eating soup, drinking from a glass; Table 3). Therefore these are very suitable to estimate real tremor intensity.

Depression (of various degree) was seen in 1/3 of the patients. There was neither a correlation of tremor intensity with psychoorganic disturbances nor the presence of an endogenomorphic depressive disorder.

A more detailed analysis is given elsewhere (Auff, 1990).

Discussion

A striking proportion of our "pure" ET patients showed some singular Parkinson symptoms, although generally very slight. This may be due to the old age of many of our patients, as this is well-known in later stages of life (Critchley, 1956). This finding as well as the occurence of other tremor components than a postural tremor (especially resting tremor; Findley, 1984; Marsden, 1984) in patients with ET may be responsible for a misdiagnosis of Parkinson's disease.

1/3 of the patients had a noticeable disturbance of dexterity or motor performance not to be ascribed to the tremor alone; this disturbance has recently been reported also by two other groups and may additionally contribute to disability (Koller et al., 1986; Wood et al., 1984); it may sustain after treatment even if the tremor improves.

Questions aiming at the extent of subjective impairment in simple, clearly defined activities of daily living like writing, eating and drinking, dressing etc. make it easy to estimate the objective dimension of characteristically dis-

Table 3. Correlation of results from the motor performance test ("Motorische Lei-stungsserie nach Schoppe", various items) and various self-rating scales (*BFS* Befind-lichkeitsskala v.Zerssen, *BL* Beschwerdenliste v.Zerssen, *SDS* Self-rating Depression Scale Zung, *VAS* visual analogue scale)

	Steadiness	Tapping	Line following	Aiming
BFS	−	−	−	−
BL	−	−	−	−
SDS	−	−	−	−
VAS eating	+	−	+	+
VAS soup	+	+	−	+
VAS drinking	−	+	+	+
VAS writing	+	+	+	+
VAS signature	+	+	+	+
VAS dressing	+	−	−	+
VAS hygiene	(+)	−	+	+
VAS working	+	(+)	+	−
VAS hobbies	−	−	+	+

+ statistically significant correlation (p < 0.05) for both hands, (+) only for one hand, − n.s.

turbed tasks correctly in essential tremor patients, whereas on the other hand general questions (e.g. about problems with working, hobbies etc.) do not give conclusive information on the tremor intensity. Naturally such statements are less useful in patients with higher psycho-organic disturbances. Our results are important for investigations of the course of the disease and also long-term therapy studies of patients with ET. If one wants to get reliable data of the tremor intensity despite the well-known and pronounced intraindividual fluctuations, one has to take this into consideration. A number of self-rating scales for subjective impairment or depression, like BL-v. Zerssen, BF-S-v. Zerssen, Zung-Depression-scale and also questions directed at complex tasks are not useful: they do not correlate with the objective measuring methods. On the other side the MLS, the clinical neurological investigation, a writing (or drawing) test and also the patients' statements of their problems with clearly defined, simple tasks are of great value. If there are no other objective measurements like accelerometric measurement and even beside these, one should use these items for judging tremor intensity in long-term therapy studies and also in the course of the disease.

References

Auff E (1990) Der essentielle Tremor. Zusammenhänge und Diskrepanzen zwischen subjektiver Beeinträchtigung und objektiven Meßergebnissen. Facultas, Wien
Critchley M (1956) Neurologic changes in the aged. J Chron Dis 3:459–477
Findley LJ (1984) Essential tremor: introductory remarks. In: Findley LJ, Capildeo R (eds) Movement disorders: tremor. Macmillan, London, pp 207–209

Koller WC, Royse VL (1985) Time course of a single oral dose of propranolol in essential tremor. Neurology 35: 1494–1498

Koller W, Biary N, Cone S (1986) Disability in essential tremor: effect of treatment. Neurology 36: 1001–1004

Marsden CD (1984) Origins of normal and pathological tremor. In: Findley LJ, Capildeo R (eds) Movement disorders: tremor. Macmillan, London, pp 37–84

Rautakorpi I, Takala J, Marttila RJ, Sievers K, Rinne UK (1982) Essential tremor in a Finnish population. Acta Neurol Scand 66: 58–67

Rautakorpi I, Marttila RJ, Rinne UK (1984) Epidemiology of essential tremor. In: Findley LJ, Capildeo R (eds) Movement disorders: tremor. Macmillan, London, pp 211–218

Sutherland JM, Edwards VE, Eadie MJ (1975) Essential (hereditary or senile) tremor. Med J Aust ii: 44–47

Wood H, Miska R, Nausieda PA (1984) Drug responsiveness and disability in essential tremor. Neurology (Cleveland) 34 [Suppl 1]: 88

Authors' address: Univ. Doz. Dr. E. Auff, Neurologische Universitätsklinik, Lazarettgasse 14, A-1090 Wien, Austria

J Neural Transm (1991) [Suppl] 33: 111–114
© by Springer-Verlag 1991

Sympathetic vascular function in patients with central dysautonomia

L. Santambrogio, G. Bellomo, M. Mercuri, R. Paltriccia,G. Ciuffetti, and **E. Mannarino**

Institute of Clinica Medica II, University of Perugia Medical School,
Policlinico Monteluce, Perugia, Italy

Summary. This study used digital photoplethysmography (d-FPG) to investigate alterations in skin blood flow after exposure to cold as well as the post-prandial blood pressure pattern to assess how the sympathetic branch of autonomic nervous system (ANS) functioned in 31 patients with cerebral dysautonomia and in 27 healthy controls. d-FPG was carried out on all ten fingers in basal conditions and after exposure to ice-cold water (4–5°C). Amplitude, crest time and inclination time were used to calculate the alterations induced by the cold pressor test. After a standard lunch blood pressure was monitored every 20 minutes using a fully automatic non-invasive sphygmanometer. Unlike the controls the amplitude of the photopletysmographic wave increased in all patients except 2; crest time and inclination time decreased in all except 3; post-prandial diastolic and systolic blood pressure levels fell markedly in all but 3. Blood vessel smooth muscle tone is disturbed in patients with ANS failure because dysautonomia may permit the action of vasodilating substances to predominate. The post-prandial blood pressure pattern and the d-FPG used in conjunction with a cold pressor test are useful tools in the non-invasive investigation of ANS function.

Introduction

The sympathetic branch of the autonomic nervous system(ANS) is of paramount importance in the regulation of vascular smooth muscle tone, especially in the skin (Clement, 1979), where it mediates constriction following a variety of stimuli the most important of which is probably a fall in the external temperature (Rooke et al., 1987). Such a vascular reaction is secondary to the release of norepinephrine by sympathetic nerve endings (Rowell, 1977) and appears to be potentiated by a cold-induced increase in the sensitivity of vascular smooth muscle α-adrenoceptors to catecholamines (Vanhoutte et al., 1985). The sympathetic ANS is also important for the vascular adaptation to potentially hypotensive stimuli such as standing or the meal-induced release of vasodilating substances by the gastroenteric tract (Rowe et al., 1981). In the present study we have investigated the change of skin blood flow following exposure to cold using digital photoplethysmography (d-FPG), and the postprandial pattern of blood pressure as measures of sympathetic ANS function in a group of patients with documented central dysautonomia, compared to healthy age-matched controls.

Patients and methods

A group of 31 patients with dysautonomia aged 54.1 ± 6.4 years (17 men and 14 women, 26 with idopathic Parkinson's disease and 5 with the Shy-Drager's syndrome) and a group of 27 healthy controls aged 51.8 ± 5.7 years (14 men and 13 women) participated in the study after being fully informed about the procedures and giving their consent. All the patients with diabetes mellitus, abnormal liver or renal function or hypertension, as well as patients with documented peripheral vascular disease were excluded from the study. Seventeen of the patients were treated with levodopa, six with levodopa and orphenadrine, three with deprenyl, and the remaining six were not taking any medications; ANS function was evaluated using the following tests, performed as described elsewhere (Ewing et al., 1980; Ewing and Cearke, 1982): R-R variations with deep-breathing, Valsalva ratio, blood pressure and heart rate responses to standing and sustained handgrip. The patients were considered as having dysautonomia if at least three of the tests yielded abnormal results. On the day of the study with d-FPG (performed between 8 and 10 a.m.) the subjects were kept for at least thirty minutes in the examination lab at a 22°–23° temperature; all of them had refrained from smoking for at least 24 hours; d-FPG was carried out on all ten fingers of the hands, in basal conditions, and immediately thereafter after keeping both hands in ice-cold water (4°–5°), again on all ten fingers; the following parameters (average of all the fingers) were computed: amplitude, crest time, inclination time; the change induced by the cold pressor test was then calculated. All the d-FPG readings were performed using a Stereodop 448.S instrument (Ultrasomed, Milan, Italy). The postprandial blood pressure adaptation was investigated two days later, using a standard lunch (12 kcal/kg ideal body weight, 65% carbohydrates, 20% fats, 15% proteins) administered at 12 a.m.; blood pressure was monitored every 20 min using a fully automatic non-invasive sphygmomanometer (BP 203Y, Behring Linea Strumentazione, Italy) as previously described (Bellome et al., 1988); maximal blood pressure decrement was calculated subtracting the nadir from the baseline represented by the average of all preprandial values. Statistical analysis was carried out using the Student's t test for independent samples.

Results

The results of conventional autonomic function tests are shown in Table 1; the data concerning d-FPG parameters and postprandial blood pressure changes are summarized in Table 2. The amplitude of the photoplethysmographic wave was increased in all but two of the patients after the cold pressor test, whereas it tended to decrease in the totality of the controls; immersion of the hands in ice-cold water also provoked a decrease of the crest time and inclination time in all but three of the patients with dysautonomia; conversely an opposite behaviour was displayed by all the controls, in whom such parameters were increased or remained unchanged after the cold pressor test. Both systolic and diastolic blood pressure decreased markedly after the meal in the patients with ANS failure, reaching a nadir between 60

Table 1. Results of conventional non-invasive autonomic function tests (mean ± SD, *p < 0.05)

	Dysautonomic	Controls
Valsalva ratio	1.04 ± 0.13*	1.66 ± 0.2
Deep breathing (beats/min)	7.7 ± 1.2*	18.0 ± 3.5
Orthostatic reaction:		
Heart rate (beats/min)	11.9 ± 2.6*	17.6 ± 5.8
Systolic BP (mmHg)	−37.7 ± 6.8*	−8.7 ± 5.3
Diastolic BP (mmHg)	−22.8 ± 4.8*	−5.5 ± 4.0
Reaction to sustained handgrip:		
Heart rate (beats/min)	5.3 ± 2.8*	14.7 ± 6.0
Systolic BP (mmHg)	6.0 ± 3.3*	28.7 ± 9.2
Diastolic BP (mmHg)	4.2 ± 2.4*	17.4 ± 5.6

and 100 minutes after eating. In three of the patients the blood pressure pattern was comparable to that of the control group.

Discussion

The results of our study demonstrate the presence of a clinically evident disturbance in blood vessel smooth muscle tone regulation in patients with ANS failure; in fact following the release of vasodilating substances induced by meal, the defective sympathetic ANS is not able to prevent a marked decrease in blood pressure, as also shown by others in patients with Parkinson's disease (Micieli et al., 1987). As far as skin blood flow is concerned, the analysis of d-FPG parameters show that, as expected, the digital arteries of normal subjects tend to constrict after exposure to cold; this does not appear to be the case when the vessels of dysautonomic patients are examined; in fact these tend to react to cold with a paradox vasodilation. The exact cause of such an abnormal reaction is not clear yet, however, we can speculate that

Table 2. Analysis of d-FPG parameters (expressed as change after the cold pressor test) and postprandial blood pressure variations (mean ± SD, *p < 0.05)

	Dysautonomic	Controls
d-FPG parameters:		
Amplitude (cm)	0.34 ± 0.24*	−0.59 ± 0.24
Crest time (%)	−1.2 ± 0.6*	5.7 ± 2.4
Inclination time (msec.)	−32.0 ± 8.9*	124.8 ± 49.7
Postprandial blood pressure changes:		
Systolic BP (mmHg)	−40.8 ± 10.6*	−9.1 ± 3.5
Diastolic BP (mmHg)	−25.3 ± 8.9*	−5.2 ± 3.3

in the absence of a functioning sympathetic ANS, the action of vasodilating substances such as bradykinins, the release of which is known to be triggered by painful stimuli such as intense cold (Roddie, 1983), predominates. Impaired neurogenic microvascular response has also been shown to occur in patients with diabetes mellitus and autonomic neuropathy (Parkhouse and Le Queshe, 1988); in the present study we have demonstrated its presence in patients with central ANS failure free of clinically evident peripheral vascular disease, indicating that ANS dysfunction alone is capable of severely altering vascular reactivity. In conclusion, we believe that the study of postprandial blood pressure reactions and the analysis of digital blood flow by d-FPG combined with a cold pressor test, are useful complements to the battery of tests used for the non-invasive investigation of ANS function in humans.

Acknowledgement

The authors would like to thank Dr. G. A. Boyd for her invaluable assistance.

References

Clement DL (1979) J Cardiovasc Surg 20: 291–342
Rooke TW, Hollier LH, Osmundson PJ (1987) Angiology 39: 400–409
Rowell LB (1977) J Invest Dermatol 69: 154–166
Vanhoutte PM, Cooke JP, Lindblad LE (1985) Clin Sci 68: 1215–1235
Rowe JW, Young JB, Minaker KL, Stevens AL, Pallotta J, Landsberg L (1981) Diabetes 30: 219–225
Ewing DJ, Campbell IW, Clarke BF (1980) Ann Intern Med 92: 308–311
Ewing DJ, Clarke BF (1982) Br Med J 285: 9116–9118
Bellomo G, Santucci S, Aisa G, Parnetti L (1988) Gerontology 34: 311–314
Micieli G, Martignoni E, Cavallini A, Sandrini G, Nappi G (1987) Neurology 37: 386–393
Roddie EC (1983) In: The cardiovascular system-peripheral circulation and organ blood flow, vol 3. American Physiological Society, Bethesda Md
Parkhouse N, Le Quesne PM (1988) N Engl J Med 318: 1306–1309

Authors' address: Dr. L. Santambrogio, II Department of Internal Medicine, University of Perugia, Policlinico Monteluce, I-06100 Perugia, Italy

J Neural Transm (1991) [Suppl] 33: 115–118
© by Springer-Verlag 1991

Epidemiology and out-patient care in Parkinson's disease — Results from a pilot-study in Northern Germany (Schleswig-Holstein)

P. Vieregge[1], J. Kleinhenz[1], H. Fassl[2], J. Jörg[3], and D. Kömpf[1]

[1]Klinik für Neurologie and [2] Institut für Medizinische Statistik und Dokumentation, Medizinische Universität zu Lübeck, and [3]Neurologische Klinik, Klinikum Barmen, Wuppertal, Federal Republic of Germany

Summary. An attempt was made to estimate prevalence on Parkinsonism using consultation rates in physicians' practices in a two-step, one-phasic pilot study in a rural-urban area of Northern Germany. Though participation of physicians was low, reported rates for Parkinsonian patients (183/100 000) were in the range yielded in comparable areas, but by different methodology. Only 64% of a large subsample of patients used L-Dopa preparations for therapy. Confirming previous results the objective records of doctors and the subjective notices of patients regarding disease state showed considerable differences in several important items.

Introduction

The unknown cause(s) of idiopathic Parkinson's disease (IPD) and the discussion on (possibly environmental) toxic aetiological factors have raised an increased interest in epidemiological studies of IPD (Tanner, 1989; Kleinhenz et al., 1990). Reported prevalence rates differ mainly due to different methodology (Kleinhenz et al., 1990). The uniform application of the door-to-door method which is independent of the local health care system, has yielded lower prevalence rates of IPD in China and Japan compared to countries in the Western hemisphere (Anderson et al., 1982; Schoenberg, 1987). Given the actual health care system and the current problems of jurisdiction in personal data transfer a door-to-door study covering all household members of a distinct area seems impossible to do at present in the Federal Republic of Germany. We therefore chose physician's practices as one means to obtain an estimate of IPD prevalence according consulting rates in an area of rural and urban Northern Germany.

Methods

A two-step, one-phasic pilot study was conducted in the Hansestadt Lübeck, the Kreis Stormarn and the Kreis Herzogtum Lauenburg between March 1st and June 30th, 1987 in 387 practices of general practitioners, internal specialists, neurologists and psychiatrists, and in three hospital ambulatories. Standardized questionnaires with appropriate items regarding signs and symptoms, diagnosis and treatment of IPD were distributed by mail to 210 physicians (mailing group) and personally to 180 physicians (personal group). A questionnaire for each patient to be filled up by himself was distributed as well. 50 physicians of the mailing group were reminded by phone in the midth of the evaluation period to participate in the study and were asked about their IPD patients personally. At the end of the study 111 physicians of the personal group could be contacted on their IPD patients. Apart from several methodological aspects to be evaluated (Kleinhenz et al., 1990), the results were compared according to different specialist groups in the physicians (neurologists vs. non-neurologists) and to the judgements made by the physicians and their patients with regard to various aspects of the disease.

Results

41 physicians of the mailing group (20%) participated in the study, whereas 111 physicians (62%) of the personal group supplied the respective informations. The total number of patients with IPD given for the investigation period was 1018, which gave the following prevalence estimates (per 100000) according to the population of the three districts:

— Hansestadt Lübeck	188
— Kreis Stormarn	206
— Kreis Herzogtum Lauenburg	147
— Total	183

Questionnaires about 195 patients were received from eight neurologists (144 patients) and eight non-neurologists (51 patients). 132 of these 195 patients had filled up the corresponding patient questionnaire. The peak age of this group (189 of 195 patients; no data in 6 questionnaires) was in the eighth decade of life (Fig. 1) with a preponderance of women in all age groups over 70. 16% of these patients had at least on hospital admission because of IPD.

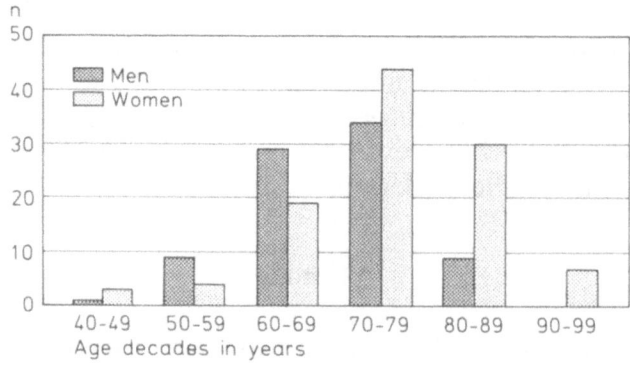

Fig. 1. Age distribution in 189 of 195 patients studied by questionnaire

Neurologists treated fewer patients over age 79 (14%) compared to non-neurologists (53%) and fewer women (45% vs. 92%). Arteriosclerotic aetiology was assumed by neurologists at a rate of 30% in their patients, by non-neurologists at 81%.

64% of the 195 patients received L-Dopa drugs alone or in combinations. Comparison between items given by doctors and their respective patients revealed a substantial agreement about drugs prescribed and taken (98%), whereas symptom items were more different: 21% of the patients denied depression, though it was marked by their doctors. 35% of the patients denied "stiffness", though rigidity was noted by their doctor. 14% of the patients denied tremor, though it was marked by their doctor. 6% of the patients gave a disease duration of more than two years prior to diagnosis.

Discussion

This first german pilot study on IPD epidemiology revealed several methodological difficulties and limitations with regard to prevalence rates: There was no proof of uniform diagnostic criteria, though the main features of IPD were items to be marked on the questionnaire. This applied to Parkinsonian syndromes of other than idiopathic aetiology as well as to the differential diagnosis of senile tremor. Migration of patients could not be traced, uneven distribution of accessable practices especially in the rural areas precludes further conclusions about regional prevalence variations. The rates are at present more likely consultation rates. Undiagnosed cases may have been missed, though lack of diagnosis after more than two years disease duration was rare. The age structure of the districts had influenced the results; for instance, the proportion of people over 65 years is higher in Lübeck (20%) than on average in the Federal Republic of Germany (15.3%) (Statistisches Landesamt Schleswig-Holstein, 1989). Sex distribution is influenced by the higher proportion of females in all three districts: Hospital studies revealed a preponderance of males (Hoehn and Yahr, 1967; Koller et al., 1986).

In the present study we did not look for further data sources (e. s. insurances, drug sales, homes for the aged). Despite this, our result of 183 patients per 100000 is in the range of what has been found in community studies of comparable areas: Aberdeen 164 (Mutch et al., 1986), Northampton 108 (Sutcliffe et al., 1985). As only 64% of our patients were treated with L-Dopa drugs, a prevalence estimation which only relies on drug sales may result in falsely low prevalence figures (Kleinhenz et al., 1990; de Pedro Cuesta, 1987).

The low proportion of hospital admissions due to IPD in our study indicates that (1) epidemiological conclusions about the problem of IPD in relation to aging and (2) clinical long-term studies, which rely predominantly on observations of regional centers for neurological or even movement disorders, may result in either inadaequate conclusions or worse prognostic indices than really present: Patients with uncomplicated and mild disease course as well as with senile-onset may be missed, because state and age are not thought to warrant specialist consultation.

From the comparison between neurologist and non-neurologist observations it is apparent that the arteriosclerotic aetiology in IPD seems to be overestimated: it does not seem to have reached non-specialist medical knowledge that arteriosclerosis today is regarded of minor aetiological importance (Jellinger, 1986). Another point of interest are the discrepancies in the perception of subjective symptoms and objective signs between doctors and their patients. These have, at fairly the same range, already been noted in the early study of Kessler (Kessler, 1972). Especially the inherent slowing of emotion and cognition in IPD and its phenomenological similarity to depressive disorders may lead to confound the patients's psychomotor behaviour with his real mood and affect (Gotham et al., 1986).

Acknowledgement

The help in conducting this study by Schering AG Berlin is kindly acknowledged.

References

Anderson DW, Schoenberg BS, Haerér AF (1982) Racial differentials in the prevalence of major neurological disorders. Neuroepidemiology 1: 17–30

Gotham AM, Brown RG, Marsden CD (1986) Depression in Parkinson's disease: a quantitative and qualitative analysis. J Neurol Neurosurg Psychiatry 49: 381–389

Hoehn MM, Yahr MD (1967) Parkinsonism: onset, progression, and mortality. Neurology 17: 427–442

Jellinger K (1986) Overview of morphological changes in Parkinson's disease. Adv Neurol 45: 1–18

Kessler II (1972) Epidemiological studies of Parkinson's disease. II. A hospital-based survey. Am J Epidemiol 95: 308–318

Kleinhenz J, Vieregge P, Fassl H, Jörg J (1990) Prävalenz des Morbus Parkinson in der Bundesrepublik Deutschland — Sind Praxiserhebungen ein geeignetes Erhebungsinstrument? Öffentl Gesundheitswesen 52: 181–190

Koller W, O'Hara R, Weiner W, Lang A, Nutt J, Agid Y, Bonnet AM, Jankovic J (1986) Relationship of aging to Parkinson's disease. Adv Neurol 45: 317–321

Mutch WJ, Dingwall-Fordyce I, Downie AW, Paterson JG, Roy SK (1986) Parkinson's disease in a Scottish city. Br Med J 292: 534–536

de Pedro Cuesta J (1987) Studies on the prevalence of paralysis agitans by tracer methodology. Acta Neurol Scand 75 [Suppl 112]: 7–106

Schoenberg BS (1987) Environmental risk factors for Parkinson's disease: the epidemiological evidence. Can J Neurol Sci 14: 407–413

Statistisches Landesamt Schleswig-Holstein (1989) Ergebnisse der Volkszählung '87. Kiel

Sutcliffe RCG, Prior R, Mawby B, McQuillan WJ (1985) Parkinson's disease in the district of Northampton Health Authority, United Kingdom. A study of prevalence and disability. Acta Neurol Scand 72: 363–379

Tanner CM (1989) The role of environmental toxins in the etiology of Parkinson's disease. Trends Neurosci 12: 49–54

Authors' address: Dr. P. Vieregge, Klinik für Neurologie, Medizinische Universität, Ratzeburger Allee 160, D-W-2400 Lübeck 1, Federal Republic of Germany

J Neural Transm (1991) [Suppl] 33: 119–124
© by Springer-Verlag 1991

Parkinsonian features in advanced Down's syndrome

P. Vieregge[1], **G. Ziemens**[2], **A. Piosinski**[3], **M. Freudenberg**[4], and **D. Kömpf**[1]

[1]Klinik für Neurologie, Medizinische Universität zu Lübeck, [2]Landeskrankenhaus Heiligenhafen, [3]Psychiatrisches Krankenhaus Rickling, and [4]Landeskrankenhaus Neustadt/Holst., Federal Republic of Germany

Summary. The results of a clinical study on extrapyramidal, frontal release, and other neurological signs in 54 demented and non-demented patients with Down's syndrome (DS) are presented. Fourteen patients were demented, five of them showed extrapyramidal signs, mainly of the rigid-hypokinetic spectrum and rather similar to Parkinsonian features in advanced Alzheimer's disease (AD). None of the non-demented patients had Parkinsonian signs. The mean ages of the demented DS patients with extrapyramidal signs was significantly higher than that of the patients without the respective signs. Frontal release signs were present in demented and nondemented patients. From a questionnaire neither a raised proportion of early- or senile-onset dementia nor of Parkinsonism among first- and second-degree relatives of DS patients could be traced. Parkinsonian signs seem to be present at a lower frequency in DS than in advanced AD. A speculative hypothesis about a gene dosage effect of Cu/Zn-superoxide dismutase in preventing toxic radical formation in the substantia nigra of DS patients is presented.

Introduction

Several lines of evidence support a close association between Alzheimer's disease (AD) and Downs's syndrome (DS): Neuropathological changes characteristic of AD are found in the brains of almost all DS individuals dying after age 35 (Wisniewski et al., 1985a). Age-related abnormalities in adaptive and social behaviour (Miniszek, 1983), neurological status (Loesch-Mdzewska, 1968; Wisniewski et al., 1978, 1985b; Lott and Lai, 1982), and EEG rhythms (Veall, 1974) have been described in DS, demonstrating at least subtle deterioration with advancing age and similar to a pattern found in AD.

Despite the morphological similarities the genetic relationship between DS and AD remains unclear: An increased number of presenile dementia in relatives of DS probands was reported (Yatham et al., 1988). Patients with AD of early onset and familial occurrence may have a predisposing locus on chromosome 21, while in families with senile onset such linkage could not be substantiated (Goate et al., 1989).

Patients with AD of an advanced stage exhibit extrapyramidal (EP) and especially Parkinsonian signs to a variant degree (Pearce, 1974; Drachman and Stahl, 1975; Sulkawa, 1982; Koller et al., 1984; Mölsä et al., 1984; Chui et al., 1985; Mayeux et al., 1985; Leverenz and Sumi, 1986; Morris et al., 1989; Tyrell and Rossor, 1989). A recent case-control study provided evidence for a familial aggregation of AD with Parkinson's disease (PD) (Hofmann et al., 1989).

In order to further elucidate the natural history of DS patients in middle and old age, we looked for EP system involvement and other neurological signs and for evidence of familial clustering of AD and PD in families with an index case of DS.

Patients and methods

54 patients (34 males, 20 females) with DS from various institutions were diagnosed on the basis of the classic phenotype. Karyotype analysis (done wherever possible): 27 patients had "free" trisomy 21, 2 had a mosaic pattern, and one a translocation trisomy.

The medical history of each patient was taken with special attention to diseases known to occur more frequently in DS (cardiac anomalies, myeloproliferative disorders, thyroid disease). A neurological examination as detailed as possible in the individual patients used fixed criteria according the Unified Parkinsons's Disease Rating Scale (UPDRS) (Fahn et al., 1987) to evaluate the degree of hypokinesia, rigidity, resting tremor, postural abnormalities, walking pattern, pulsion phenomena etc. Special attention was given to dyskinesias, myoclonic jerks, pyramidal signs and signs of "frontal disinhibition": palmomental reflex (PMR), grasping reflex (GR), sucking and/or snouting reflex (SSR).

Dementia was assessed clinically by interview and/or by assessment of those caring for the patients. 35 patients had EEG examination, 9 patients cranial CT. A mailed questionnaire with items regarding the occurrence of dementia and parkinsonism in first- and second-degree relatives was obtained from the families of 38 patients.

Results

Table 1 shows the mean ages of the patient sample with regard to dementia, extrapyramidal, and frontal signs. 14 DS patients (age range 27–67 years) were assessed to be demented. Their mean age was significantly higher than that of the 40 non-demented DS patients.

The mean age of 5 demented DS patients with extrapyramidal signs was significantly higher than that of the 9 demented patients without extrapyramidal involvement. The mean age of patients with palmomental reflex and of those with grasping reflex was significantly higher than that of the patients without these signs. In the extrapyramidal involvement rigid-hypokinetic signs prevailed (Table 2). Resting tremor and myoclonic jerks were not observed. No patient received L-DOPA treatment.

Other neurological signs unrelated to dementia were mainly oculomotor and motor "soft signs". Grand-mal epilepsy was present in 7 patients (two

Table 1. Number and mean age ($\pm$ SEM) of patients with Down's syndrome with dementia, extrapyramidal involvement, and signs of frontal disinhibition

Patient group	Number (%)		Mean age (years)
Whole sample	54		44.4 + 9.9
Male patients	34	(63)	44.4 ± 10.0
Female patients	20	(37)	44.9 ± 10.0 NS
Non-demented	40	(74)	42.2 + 8.6 ***
Demented	14	(26)	50.9 + 10.9
Extrapyramidal			
Signs present	5/14	(36)	59.8 ± 6.7 **
absent	9/14	(64)	45.8 ± 9.7
PMR present	30	(56)	48.7 + 8.1 ***
absent	24	(44)	39.0 + 9.5
GR present	5	(9)	54.2 ± 11.1 *
absent	49	(91)	43.4 ± 9.4
SSR present	4	(7)	47.3 ± 3.8 NS
absent	50	(93)	44.2 + 10.2

*, **, *** denotes $p < 0.05$, < 0.01, < 0.001 respectively (Student's t-test); *NS* not significant; *PMR* palmomental reflex; *GR* grasping reflex; *SSR* sucking and snouting reflex

Table 2. Extrapyramidal and "frontal" signs in 14 demented patients with Down's syndrome

Sign	Number of patients (%)
Rigidity	3 (21)
Shuffling gait	4 (29)
Hypokinesia	3 (21)
Hypomimia	3 (21)
Orofacial dyskinesia	2 (14)
Palmomental reflex	9 (64)
Grasping reflex	3 (21)
Sucking/snouting reflex	2 (14)

demented). EEG was normal in 11 patients, mildly abnormal in 13, and moderately abnormal in 11. None of the patients with seizures had epileptic discharges in the interictal EEG recording. There was no relation between severity of EEG changes and signs of frontal disinhibition or dementia.

Three non-demented patients had basal ganglia calcification, but no signs of extrapyramidal involvement.

From the questionnaire no first-degree relatives with early-onset dementia (before age 65) could be detected. One mother had Parkinsonism before age 65. Among 149 first- and second-degree relatives older than 65 years two had senile dementia (mother, paternal grandfather), and two had Parkinsonism (father, paternal grandfather).

Discussion

EP signs in older DS patients (36% of the demented; 9% of all) were more frequently seen than in other reports, where such signs were not mentioned or even absent (Loesch-Mdzewska, 1968; Wisniewski et al., 1978, 1985b; Owens et al., 1971). In a most recent investigation 20% of demented DS patients of middle and old age showed Parkinsonian signs (Lai and Williams, 1989).

Two patients of that prospective investigation had Parkinsonism prior to the onset of dementia, whereas all of our DS patients with EP signs were demented. These patients were of significantly higher age compared to those demented DS patients without EP involvement. It appears therefore from our study, that EP signs are features more likely to occur late in the clinical course of demented DS patients, but that such signs do not seem to be obligate in the disease spectrum.

The overall amount of 36% of demented patients with EP signs in DS is the range also found in AD patients (Chui et al., 1985; Mayeux et al., 1985; Leverenz and Sumi, 1986; Morris et al., 1989). Others, however, have presented considerably lower (9%; Koller et al., 1984) or even higher figures for their AD patients: 62% (Pearce, 1974); 65% (Tyrell and Rossor, 1989); 92% (Mölsä et al., 1984).

The clinical feature in our study was mainly of the hypokinetic rigid spectrum. This bears similarities to Parkinsonism encountered during the course of AD (Pearce, 1974; Drachman and Stahl, 1975; Sulkava, 1982; Koller et al., 1984; Mölsä et al., 1984; Chui et al., 1985; Mayeux et al., 1985; Leverenz and Sumi, 1986; Morris et al., 1989; Tyrell and Rossor, 1989).

The SN in DS is less severely involved pathologically in young and middle age than other subcortical nuclei, which was found to be in contrast to respective AD cases (Mann et al., 1987). On the other hand, DS brains of advanced disease duration show more numerous neurofibrillar tangles than respective AD brains and mild to moderate SN pathology suggestive of Parkinsonism (Gibb et al., 1989). These findings indicate that the pathological changes in the SN and the clinical presentation of Parkinsonism seem to be phenomena occuring late in the course of DS.

Why does the SN in DS seem to be afflicted so late in the disease when compared to other brain regions and to a less severe degree than in AD? One should remember that the triplication of the chromosomal segment 21q22 is responsible for an overexpression of the gene for the Copper/Zinc-superoxide dismutase (Cu/Zn-SOD).

In Parkinson's disease hydrogen peroxide and related oxy-radicals are thought to be involved in the degeneration of dopamine neurons (Cohen, 1983). Cytosolic Cu/Zn-SOD-like activity as well as mitochondrial Mn-SOD-like activity have been found to be increased in SN of postmortem Parkinsonian brains (Marttila et al., 1988; Saggu et al., 1989). It may therefore be speculated, that in DS patients the excess of Cu/Zn-SOD serves as a protective mechanism in the SN by scavenging radicals occurring in nigral metabolism. This may result clinically in a Parkinsonian syndrome at a later stage, and, if ever, to a lesser degree than expected from comparison to AD.

In our questionnaire-based inquiry on dementia and PD signs in the families of our DS patients there was no increased rate of dementia with onset before age 65. The number of two senile-onset demented relatives among 149 persons at risk does not sharply exceed to what is exspected according to the age-specific prevalence rate for dementia over 65 years of age in the general population (0.65%) (Schoenberg et al., 1987): (2 × 100) : 149 = 1.34%. The same applies for two relatives wth senile-onset (after age 65) PD, given the similar age-specific prevalence rates as for dementia in the seventh and eighth decade of life.

The present investigation has shown, that extrapyramidal signs may be a feature of advanced DS. They were encountered only in demented DS patients. This finding parallels Alzheimer's disease, where such signs may also occur in an advanced stage, but to a greater degree. In order to evaluate, if the extrapyramidal involvement in DS may help to clarify the debated association between AD and PD, further clinico-pathological and neurochemical studies in DS patients may be useful. As delineated above, such studies may gain additional interest considering the role of free-radical formation and its prevention in the pathophysiology of idiopathic Parkinson's disease.

Note added in proof

Since submission of this study a single case report of "Levodopa-responsive Parkinsonism in a patient with Down's syndrome" has been published by Singer C, Sanchez-Ramos J, Weiner WJ in European Neurology (1990) 30: 247–248.

References

Chui HC, Teng EL, Henderson VW, Moy AC (1985) Clinical subtypes of dementia of the Alzheimer type. Neurology 35: 1544–1550

Cohen G (1983) The pathobiology of Parkinson's disease: biochemical aspects of dopamine neuron senescence. J Neural Transm [Suppl 19]: 89–103

Drachman DA, Stahl S (1975) Extrapyramidal dementia and levodopa. Lancet i: 809

Fahn S, Elton RL, and members of the UPDRS development committee Unified Parkinson's disease rating scale (1987). In: Fahn S, Marsden CD, Calne DB, Goldstein M (eds) Recent developments in Parkinsons disease, vol 2. Macmillan Health Care Information, Florham Park NJ, pp 153–167

Gibb WRG, Mountjoy CQ, Mann DMA, Lees AJ (1989) The substantia nigra and ventral tegmental area in Alzheimer's disease and Down's syndrome. J Neurol Neurosurg Psychiatry 52: 193–200

Goate AM, Haynes AR, Owen MJ, Farrall M, James LA, Lai LYC, Mullan MJ, Roques P, Rossor MN, Williamson R, Hardy JA (1989) Predisposing locus for Alzheimer's disease on chromosome 21. Lancet i: 352–355

Hofman A, Schulte W, Tanja TA, van Duijn CM, Haaxma R, Lameris AJ, Otten VM, Saan RJ (1989) History of dementia and Parkinson's disease in 1st-degree relatives of patients with Alzheimer's disease. Neurology 39: 1589–1592

Koller WC, Wilson RS, Glatt SL, Fox JH (1984) Motor signs are infrequent in dementia of the Alzheimer type. Ann Neurol 16: 514–516

Lai F, Williams RS (1989) A prospective study of Alzheimer disease in Down syndrome. Arch Neurol 46: 849–853

Leverenz J, Sumi SM (1986) Parkinson's disease in patients with Alzheimer's disease. Arch Neurol 43: 662–664

Loesch-Mdzewska D (1968) Some aspects of the neurology of Down's syndrome. J Ment Defic Res 12: 237–246

Lott IT, Lai F (1982) Dementia in Down's syndrome: observations from a neurologic clinic. Appl Res Ment Retard 3: 233–239

Mann DMA, Yates PO, Marcyniuk B, Ravindra CR (1987) Loss of neurons from cortical and subcortical areas in Down's syndrome patients at middle age. J Neurol Sci 80: 79–89

Marttila RJ, Lorentz H, Rinne UK (1988) Oxygen toxicity protecting enzymes in Parkinson's disease. Increase of superoxide dismutase-like activity in the substantia nigra and basal nucleus. J Neurol Sci 86: 321–331

Mayeux R, Stern Y, Spanton S (1985) Heterogeneity in dementia of the Alzheimer type: evidence of subgroups. Neurology 35: 453–461

Miniszek NA (1983) Development of Alzheimer disease in Down's syndrome individuals. Am J Ment Defic 87: 377–385

Mölsä PK, Marttila RJ, Rinne UK (1984) Extrapyramidal signs in Alzheimer's disease. Neurology 34: 1114–1116

Morris JC, Drazner M, Fulling K, Grant EA, Goldring J (1989) Clinical and pathological aspects of Parkinsonism in Alzheimer's disease. Arch Neurol 46: 651–657

Owens D, Dawson JC, Losin S (1971) Alzheimer's disease in Down's syndrome. Am J Ment Defic 75: 606–612

Pearce J (1974) The extrapyramidal disorder of Alzheimer's disease. Eur Neurol 12: 94–103

Saggu H, Cooksey J, Dexter D, Wells FR, Lees A, Jenner P, Marsden CD (1989) A selective increase in particulate superoxide dismutase activity in Parkinsonian substantia nigra. J Neurochem 53: 692–697

Schoenberg BS, Kokmen E, Okazaki H (1987) Alzheimer's disease and other dementing illnesses in a defined United States population: incidence rates and clinical features. Ann Neurol 22: 724–729

Sulkava R (1982) Alzheimer's disease and senile dementia of Alzheimer type. Acta Neurol Scand 65: 636–650

Tyrell PJ, Rossor MN (1989) Extrapyramidal signs in dementia of Alzheimer type. Lancet ii: 920

Veall RM (1974) The prevalence of epilepsy among mongols related to age. J Ment Defic Res 18: 99–106

Wisniewski KE, Howe J, Williams CG, Wisniewski HM (1978) Precocious aging and dementia in patients with Down's syndrome. Biol Psychiatry 13: 619–627

Wisniewski KE, Wisniewski HM, Wen GY (1985b) Occurrence of neuropathological changes and dementia of Alzheimer's disease in Down's syndrome. Ann Neurol 17: 278–282

Wisniewski KE, Dalton AJ, CrapperMcLachlan DR, Wen GY, Wisniewski HM (1985b) Alzheimer's disease in Down's syndrome: clinico-pathologic studies. Neurology 35: 957–961

Yatham LN, McHale PA, Kinsella A (1988) Down's syndrome and its association with Alzheimer's disease. Acta Psychiatr Scand 77: 38–41

Authors' address: Dr. P. Vieregge, Klinik für Neurologie, Medizinische Universität zu Lübeck, Ratzeburger Allee 160, D-W-2400 Lübeck 1, Federal Republic of Germany

J Neural Transm (1991) [Suppl] 33: 125–132
© by Springer-Verlag 1991

Effects of treatment with trihexyphenidyl on cognitive function in early Parkinson's disease

L. Schelosky[1], **Th. Benke**[2], and **W. H. Poewe**[1]

[1]Department of Neurology, UKRV, Berlin, Federal Republic of Germany, and
[2]University Clinic for Neurology, Innsbruck, Austria

Summary.   13 patients with PD of recent onset underwent a series of neuro-psychological tests for frontal lobe associated functions (Sternberg paradigm, WCST, CVLT) before and during treatment with Artane.® Test results at baseline were not significantly different from those of an age-matched control group ($n = 13$). Retesting after a mean of 2 weeks' treatment with trihexyphenidyl revealed only slight impairment in CVLT while performance on the other tests remained unchanged.

Introduction

The prevalence of dementia in Parkinson's disease has been estimated to be around 20%, twice the figure given for the general population (Gibb, 1989). Possible morphological substrates include concomitant Alzheimer pathology, cell loss in the Nucleus Basalis of Meynert and diffuse cortical Lewy body degeneration (Gibb, 1989; Hakim and Mathieson, 1979; Whitehouse et al., 1988). Neurochemically both Alzheimer type cortical degeneration and neuronal loss in the Nucleus Basalis are associated with frontal cholinergic deficiency. Anticholinergic treatment of Parkinson's disease may therefore induce or aggravate cognitive impairment and several studies have demonstrated this in patients receiving anticholinergics as add-on treatment with levodopa (Sadeh et al., 1982; Dubois et al., 1987; Miller et al., 1987; Koller, 1984). There are however few studies of the effects of anticholinergic therapy on specific cognitive domaines in early Parkinson's disease. The present study was performed to investigate the effects of trihexyphenidyl on cognitive functions associated with the frontal lobes in patients with Parkinson's disease of recent onset, the majority of whom had not received previous antiparkinson treatment.

Subjects and methods

13 subjects with idiopathic Parkinson's disease were included in the study. 3 patients were on oral levodopa, one patient was treated with the ergot dopamine agonist CQA

Table 1

	Patients n = 13	Controls n = 13	p-value
Age	63,23 (46–77)	56,8 (42–74)	n.s.
Sex (f:m)	9:4	8:5	n.s. [+]
Years of education	10,38	9,46	n.s.
Duration of illness (years)	2,3 (1–5)		
H + Y stage	2,1 (1–3)		
UPDRS III at baseline	18,72 (8–27)		
Treatment details:			
de novo	n = 8		
previously treated	n = 5		
L-Dopa	n = 3		
duration (years)	1.07 (1–5)		
dosage (mg)	123.1 (0–400)		
Dopamine agonists	n = 4		

[+] Chisquare test

206–291 and one with deprenyl. 8 patients had never before received any antiparkinson treatment ("de novo") (details see Table 1). Trihexyphenidyl was introduced into therapy to control the tremor.

Subjects were assessed before the start of anticholinergic therapy and after a mean of two weeks of treatment (2–4) with constant dosage of 6 mg trihexyphenidyl daily.

13 age and sex matched volunteers without evidence of CNS disease served as controls for WCST and the computerised RT measures at baseline (see Table 1).

Neuropsychological test procedures were designed to assess functions of intelligence, memory and learning ability, set shifting aptitude and cognitive processing speed and included the following:

Intelligence testing

Verbal IQ (VIQ) was measured by use of verbal subtests of the HAWIE, a german equivalent of the Wechsler Adult Intelligence Scale (information, digit span, vocabulary, arithmetic, comprehension, similarities), and performance IQ (PIQ) was calculated from the non-verbal tests (picture completion, block design, picture arrangement, object assembly, digit-symbol test). The sum of all scores yielded the general IQ (GIQ). IQ testing of controls comprised the WIP, a shortened HAWIE version.

Memory

A german version of the Wechsler Memory Scale (WMS) was used to test verbal memory functions (information, orientation, story reproduction, digit span, associative learning). Figural memory was assessed by use of the Benton Visual Retention Test (immediate reproduction of geometric items), short term memory by use of Knox cubes and Corsi block tapping (version according to Milner, 1971) (Lezak, 1983).

The California Verbal Learning Test (CVLT) in its german version was employed to assess learning ability. This test consists of the fivefold repetition of 16 words as typical representatives of four common categories, interrupted by immediate free recall of the presented words. Following the presentation of a distractor list of 16 unrelated items subjects are asked to recall as many of the words of the first list as possible, first in free recall, then following semantic cues. Long delay free and cued recall are assessed after twenty minutes filled with nonverbal test procedures, when recognition of the presented words is also checked (Delis et al., 1986; Pope, 1987).

"Frontal lobe tests"

Semantic memory and associative verbal production were tested by time limited (60 seconds) word list generation for categories (supermarket, animals) and letters (s). As a further investigation of frontal lobe function the Wisconsin Card Sorting Test (WCST) was administered at baseline only (Heaton, 1980).

Cognitive processing speed

The changes of mental processing time were checked with a modified version of the Sternberg paradigm (Sternberg, 1966; Wilson et al., 1980) measuring high speed memory scanning after presentation of sets of digits. A computerised version of the Digit Symbol Substitution Task (Rogers et al., 1987) was also employed. Subjects were required to respond to visual presentation of digits or symbols (standing as substitutes for distinct digits) by the same motor act (pressing one of four number keys). The difference of choice reaction time of symbol substitution task as against simple reaction time of digit presentation served as measure of cognitive processing speed.

At every test session patients were rated according to the Unified Rating Scale of Parkinsonism (UPDRS).

Statistical evaluation

Comparisons were made between patients at baseline and controls as well as between patients at baseline and after a mean of two weeks (2–4) of treatment with 6 mg trihexyphenidyl daily. Results offering parametric data were computed with the Student's t-test, nonparametric data with the Wilcoxon signed rank test.

Results

Performance at baseline compared to healthy controls

Intelligence testing revealed no difference between healthy age matched controls and parkinsonian patients. In the WCST patients showed a tendency towards reduced number of achieved categories which was not statistically significant, but produced similar amounts of perseverative and nonperseverative responses and errors as controls. Verbal fluency and WMS were within normal limits (Lezak, 1983), cognitive processing speed in the Sternberg paradigm or Digit Symbol Substitution Task were equal between controls and patients, but motor reaction times showed an increase in the parkinsonian population. With respect to the CVLT no results for age matched normal controls were available in this study, but patients seemed to

L. Schelosky et al.

Table 2. Neuropsychological test results of PD group

Trihexyphenidyl	without	with
HAWIE		
full scale IQ	98.5	
verbal IQ	101.8	
performance IQ	95.7	
WCST		
correct responses	68.7	
perseverative responses	28.7	
perseverative errors	25.6	
nonpers. errors	22.7	
categories	3.5	
WMS		
information	5	5
orientation	5	5
story reproduction	94.2	89.9
associative learning	101.3	89.9*
digit span	100	99.7
Knox cubes	5.5	5.5
Corsi's block tapping	12	11.2
BVRT	10.9	10.23
Verbal fluency (words)	50.6	50
Reaction times (msec):		
Sternberg motor	1191.1	1124.8
Sternberg cognitive	48.6	42.5
Digit symbol (no encoding)	991.6	903.9
Digit symbol (encoding)	1210.8	1170.1

* $p < 0.03$

Table 3. Neuropsychological test results (controls)

WIP	
full scale IQ	103.4
verbal IQ	102.3
performance IQ	95.8
WCST	
correct responses	67.8
perseverative responses	34.8
perseverative errors	29.3
nonperseverative errors	19.0
categories	4.4
Reaction times (msec)	
Sternberg motor	660.8
Sternberg cognitive	95.0
Digit symbol (no encoding)	773.3
Digit symbol (encoding)	1067.2

perform subnormally concerning their learning ability when compared with results of Ilmberger and colleagues (Karamat et al., 1989; Pope, 1987), the number of words learned and the number of semantic clusters prominently being affected.

Performance at baseline compared to anticholinergic treatment phase

No changes of cognitive processing speed could be seen after two weeks of anticholinergic treatment, and verbal fluency, BVRT, Knox cubes and Corsi block tapping showed no difference as did most of the WMS subtests.

Learning of word lists in the WMS subtest "associative learning" ($p < 0.03$) or CVLT was impaired during the drug period. In the CVLT patients learned 2.5 words less on average when retested under trihexyphenidyl treatment. Short and long delay free as well as cued recall reached a lower level. The total amount of semantic clusters as a measure of storing principle decreased from 8.9 to 6.8. Recognition memory remained intact (14.69 compared with 14.33 correct recognitions), only a greater difficulty in distinction wether a word had been presented or not was obvious under anticholinergic treatment (false positive 0.77 as against 2.58 words). Intrusions and perseverations remained at the same level at both test sessions (intrusions 2.1 to 3.3, perseverations 4.1 to 3.3). Although results reached significance only for values of the first cued recall following the distractor list

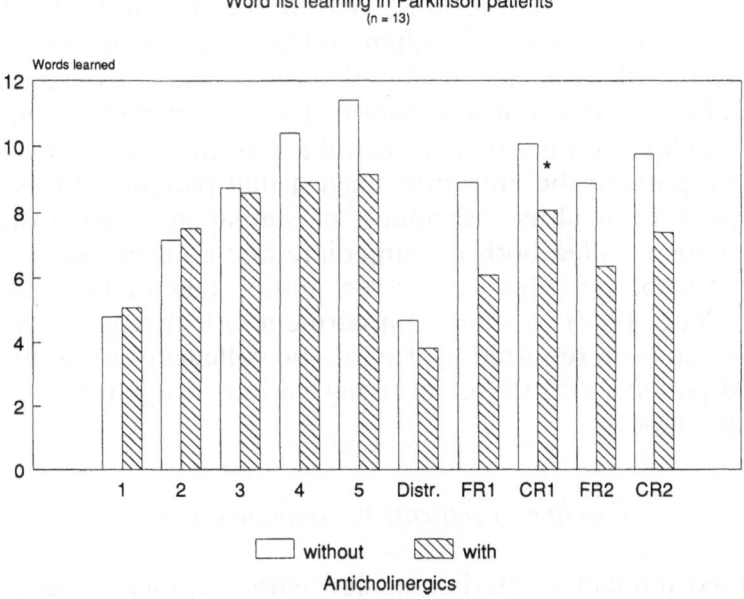

Fig. 1. Word list learning as assessed by means of CVLT. 1–5: learning during five repetitive pres- entations. *Distr* learning of distractor list. *FR1/CR1* free/cued recall of first list immediately after distractor list. *FR2/CRS* free/cued recall of first list after 20 minutes; *$p < 0.05$

(p < 0.05) they showed a clear tendency towards impaired learning, especially long term storage, of verbal material (see Fig. 1).

Score of the UPDRS showed a significant improvement of motor function with a pronounced decrease of tremor during anticholinergic treatment (total UPDRS III values 18.8 at baseline vs. 15.8 with anticholinergic treatment, p < 0.04; UPDRS III tremor scores 3.3 vs. 2.2; p < 0.03).

Discussion

Baseline of patients vs. controls

(a) The patient population of this study showed no sign of dementia as assessed by means of the HAWIE. WCST results at baseline were also not significantly different from normal controls. This is in contrast to findings in literature, where different populations of Parkinsonian patients, treated or untreated, achieved less categories or developed a more perseverative behaviour than age matched controls (Bowen et al., 1975; Lees and Smith, 1983; Taylor et al., 1986; Starkstein et al., 1987; Gotham et al., 1988). In the present study this difficulty in set shifting was not prominent in the study group, but a tendency to a reduced number of categories could be found, a trend wich might have become significant with a greater number of subjects.

(b) Cognitive processing speed of patients with Parkinson's disease of recent onset showed no difference to the control group in this study. This is in accordance with the findings of Rogers et al. (Rogers, 1986), who detected cognitive slowing in Digit Symbol Substitution Task only in a subgroup of patients with structural brain lesions. Wilson and colleagues (Wilson et al., 1980) employed the Sternberg paradigm in Parkinsonian patients of different age and only the older group exhibited reduced mental processing speed. "Bradyphrenia" was thus not a feature of the patient population studied.

(c) Although there were no control values for the CVLT available in this study, some reports in the literature suggest that patients in this series performed worse than healthy individuals of similar age with respect to the number of items recalled both in immediate and delayed recall tasks while there seemed to be no deficit in the recognition task of this test (Karamat et al., 1989; Pope, 1987). A similar impairment of learning ability in patients with PD has also been reported by Taylor and colleagues for a mixed sample of untreated patients and those receiving various drugs using the RAVLT (Taylor et al., 1986).

Baseline of patients vs. treatment phase

Only a limited number of studies dealt with memory impairment under anticholinergic treatment in Parkinson's disease. According to the present results Sadeh found no impairment of the immediate but of intermediate memory (Sadeh et al., 1982). Miller tested a large number of patients and correlated drugs used and drug intake with memory performance, finding an impairment of memory (free recall and recognition of word lists) in Parkinson's disease which is aggravated by benzhexol (Miller et al., 1987). Dubois applicated subthreshold doses of scopolamine to both controls and Parkinson patients. He observed memory difficulties only under time pressure or during tests requiring an active organization of response. He mentioned that frontal dysfunction could be responsible for these effects (Dubois et al., 1987).

In the present study only subtle cognitive changes in long term memory appear after administration of the anticholinergic agent trihexyphenidyl. Atkinson and Shiffrin divided the memory into two parts (Atkinson and Shiffrin, 1971). Short term memory (STM) has a very limited capacity which best is reflected by the digit span. Its content is transient and every new environmental information entered in this system erases that previously stored. By encoding, chunking, and other methods the stored information might enter into the long term memory (LTM) where it is permanently stored. The impairment of CVLT during treatment might be due to difficulties in retrieval out of LTM. After extinguishing the stored information in STM by a distractor list significant deterioration of free recall out of LTM results as the main cognitive change during anticholinergic treatment. Koller studied de novo patients in a similar way as this study did and found the same results — impairment of LTM but not of STM (Koller, 1984).

Studies of the effects of anticholinergics on memory functions in healthy persons have yielded essentially similar results (Kopelman, 1986; McEvoy et al., 1987; Drachman, 1977) so that the present findings do not seem to provide evidence for an increased sensitivity or specific vulnerability of non-demented Parkinsonian patients towards the cognitive side effects of anticholinergics.

References

Atkinson RC, Shiffrin RM (1971) The control of short-term memory. Scientific American Offprint 538: 1–11

Bowen FP, Kamienny RS, Burns MM, Yahr MD (1975) Parkinsonism: effects of levodopa treatment on concept formation. Neurology 25: 701–704

Delis DC, Kramer J, Ober BA, Kaplan E (1986) The California verbal learning test: administration and interpretation. Preliminary Manual

Drachman DA (1977) Memory and cognitive function in man: does the cholinergic system have a specific role? Neurology 27: 783–790

Dubois B, Danzé F, Pillon B, Cusimano G, Lhermitte F, Agid Y (1987) Cholinergic-dependent cognitive deficits in Parkinson's disease. Ann Neurol 22: 26–30

Gibb WRG (1989) Dementia and Parkinson's disease. Br J Psychiatry 154: 596–614

Gotham AM, Brown RG, Marsden CD (1988) "Frontal" cognitive function in patients with Parkinson's disease "on" and "off" levodopa. Brain 111: 299–321

Hakim AM, Mathieson G (1979) Dementia in Parkinson's disease: a neuopathological study. Neurology 29: 1209–1214

Heaton RK (1980) Wisconsin card sorting test manual. Psychological Assessment Resources, Inc

Karamat E, Ilmberger J, Poewe W, Gerstenbrandt F (1990) Memory dysfunction in Parkinson patients: an analysis of cerebral learning processes (this volume)

Koller WC (1984) Disturbance of recent memory function in parkinsonian patients on anticholinergic therapy. Cortex 20: 307–311

Kopelman MD (1986) The cholinergic neurotransmitter system in human memory and dementia: a review. Quart J Exp Psychol 38A: 535–573

Lees, AJ, Smith E (1983) Cognitive deficits in the early stages of Parkinson's disease. Brain 106: 257–270

Lezak MD (1983) Memory functions. In: Lezak MD (ed) Neuropsychological assessment. Oxford University Press, New York Oxford, pp 414–467

McEvoy JP, McCue M, Spring B, Mohs RC, Lavori PW, Farr RM (1987) Effects of amantadine and Trihexiphenidyl on memory in elderly normal volunteers. Am J Psychiatry 144: 573–577

Miller E, Berrios GE, Politynsa B (1987) The adverse effect of benzhexol on memory in Parkinson's disease. Acta Neurol Scand 76: 278–282

Pope DM (1987) The California verbal learning test: performance of normal adults aged 55–91 (unpublished)

Rogers D (1986) Bradyphrenia in parkinsonism: a historical review. Psych Med 16: 257–265

Rogers D, Lees AJ, Smith E, Trimble M, Stern GM (1987) Bradyphrenia in Parkinson's disease and psychomotor retardation in depressive illness. Brain 110: 761–776

Sadeh M, Braham J, Modan M (1982) Effects of anticholinergic drugs on memory in Parkinson's disease. Arch Neurol 39: 666–667

Starkstein SE, Leiguarda R, Gershanik O, Berthier M (1987) Neuropsychological disturbances in hemi-Parkinson's disease. Neurology 37: 1762–1764

Sternberg S (1966) High-speed scanning in human memory. Science 153: 652–654

Taylor AE, Saint-Cyr JA, Lang AE (1986) Frontal lobe dysfunction in Parkinson's disease. Brain 109: 845–883

Whitehouse, PJ, Martino AM, Marcus KA, Zweig RM, Singer HS, Price DL, Kellar KJ (1988) Reductions in acetylcholine and nicotine binding in several degenerative diseases. Arch Neurol 45: 722–724

Wilson RS, Kaszniak AW, Klawans HL, Garron DC (1980) High speed memory scanning in Parkinsonism. Cortex 16: 67–72

Authors' address: Dr. W. H. Poewe, Department of Neurology, UKRV, Spandauer Damm 130, D-W-1000 Berlin 19, Federal Republic of Germany

J Neural Transm (1991) [Suppl] 33: 133–140
© by Springer-Verlag 1991

MRI in basal ganglia diseases

D. Wimberger[1,2], **L. Prayer**[2,3], **J. Kramer**[2,3], **H. Binder**[4], and **H. Imhof**[2,3]

[1]Neurological Clinic, and [2]MRI-Institute, University of Vienna, [3]Radiodiagnostic Clinic,
University of Vienna and Ludwig Boltzmann Institute of Radiological Physical Tumor
Diagnosis, [4]Neurological Hospital Mariatheresienschlössel, Vienna, Austria

Summary. 76 patients suffering from different basal ganglia diseases (28 cases with M. Parkinson, secondary parkinsonism and Parkinson diseases; 5 cases with Chorea Huntington; 5 cases with Fahr disease and 38 cases with M. Wilson) MRI featured 2 characteristical patterns:

1. abnormal deposition of minerals,
2. focal atrophies of involved organs.

Thus MRI provides with informations about:
1. differential diagnosis in clinically misleading courses,
2. stage and, as a consequence, prognosis of some diseases,
3. biochemical processes of diseases in vivo.

Introduction

Basal ganglia diseases may feature similar clinical symptoms at least during their initial periods (Flügel, 1985). Neuropathologically they have different patterns of lesions. Up to now these patterns could hardly be recognized by imaging methods. Computed tomography only shows unspecific widening of the CSF spaces, even when clinical symptoms are already pronounced (Huckmann, 1982). Further basal ganglia diseases are related with imbalances of mineral metabolisms. Changes of iron, copper or calcium metabolism could only be recognized by biochemical tests, which do not permit any conclusion on the extent of the morphological damage.

MRI is exspected to provide with subtile information about as well morphological as metabolic aspects of basal ganglia diseases. On the one hand all parts of the extrapyramidal motor system and their morphological alterations can be identified because of MRI related sharp grey / white differentiation, on the other hand deposition of ferritin or other minerals cause characteristic signal changes. Ferritin produces a local inhomogeneity in the magnetic field that accelerates spin dephasing and results in loss of signal on T2 weighted images. Degree of signal loss varies according to the concentration of makromolecular complexes and the square of the field strength (Rutledge et al., 1987).

The aim of our study was to examine whether MRI can contribute anything to differential diagnosis and therapeutical management of basal ganglia diseases.

Patients and methods

We examined 76 patients suffering from different diseases involving the basal ganglia. Due to their pre-MRI diagnoses patients are divided up into 4 groups:

Group 1: 28 patients with M. Parkinson and Parkinson diseases.
— M. Parkinson (MP) was diagnosed in patients who had at least 3 of 4 cardinal manifestations (tremor, rigidity bradykinesia and postural instability) and were dopamine responders. 13 patients, (5 men and 8 women, 45–81 years, mean 66,2 years), fullfilled these criteria. In 6 patients vascular risk factors like arterial hypertension, smoking more than 10 cigaretts a day, adipositas or hyperlipemia was found, none of them presented cerebrovascular symptoms.
— 4 patients, (1 man, 3 women, 57–76 years, mean 64,2 years) had only 1 or 2 of the mentioned cardinal manifestations and / or were dopamine non-responders. They all had vascular risk factors, 2 had a hemiparesis resulting from ischemic infarction. All of the mentioned patients suffered from parkinsonian symptoms between 5 and 21 years (mean duration 8,3 years). 2 patients, 1 man 47 years and 1 woman, 38 years developed extrapyramidal symptoms after cerebral hypoxy. For knowing the probable reason of parkinson symptoms of these patients their disease was classified as secondary Parkinsonism (SP).
— 9 patients, (2 men, 7 women, 27–63 years mean 38,4 years), presented parkinsonian symptoms too. In addition they suffered from other complaints like, for instance, cerebellar ataxia which suggested the diagnosis of a multiple system atrophy. Among this group olivo- ponto-cerebellar atrophy (OPCA) was suspected in 4 cases, spinocerebellar atrophy in 2 cases, progressive supra — nuclear palsy in 3 cases. Duration of illness in this group was between 6 month and 14 years, mean duration 7,5 years. Diagnoses of these patients are summarized as Parkinson diseases (PD).

Group 2: 5 patients, (1 man, 4 women, 38–63 years, mean 54 years), suffered from a Chorea Huntington. Diagnosis was suspected in 4 cases because of a positive family history and typical clinical symptoms. In 1 case hypokinetic rigidity was the main symptom, thus a Parkinson disease was taken into account too. Choreatic symptoms were lasting between 6 and 15 years, mean duration 9,4 years.

Group 3: In 5 patients, (5 women, 40–65 years, mean age 56 years), Fahr disease was diagnosed. Each of these patients had postoperative hypoparathyroidism. A psychoorganic syndrome was the leading clinical sign, extrapyramidal disturbances were only mild. Onset of symptoms could not be verified exactly, thyroid operation was 12 to 24 years ago, mean 17,3 years before MR examination.

Group 4: 38 patients (21 men, 17 women, 14–58 years mean 30 years) suffered from a biochemically proved M. Wilson. The diagnosis of Wilson's disease was based on neurologic and / or hepatic symptoms, presence of Kaiser — Fleischer corneal ring at slit lamp examination, decreased serum levels of coeruloplasmin, elevated copper excretion, increased liver copper content and / or family history. Duration of disease was from several month to 27 years (mean duration 10,6 years).

For MRI we used a 1,5T Magnetom 63 (Siemens). T1 weighted (TR / TE 700/15 msec) and T2 weighted (TR / TE 2500/15,30 msec) axial slices were done, in case of infratentorial symptoms a sequence in sagittal orientation was added. Slice thickness was 5 mm, regions of interest were examined in 3 mm slices.

T2 weighted hyperintensities, corresponding to microangopathy, leukoaraiosis or focal demyelinisation, may occur in all groups. Other pathologic changes which may be revealed by MRI differ from each other. Thus different imaging criteria had to be evaluated in each group:

Group 1:
— Pars compacta width, which was measured according to the criteria of Braffman et al. (1988) and compared with an age related group of normals.
— T2 weighted signal relation between Globus pallidum and Putamen
    1. intensity of Putamen > Pallidum (normal)
    2. intensity of Putamen < / = Pallidum (pathological)
— T2 weighted hyperintense lesions
— infratentorial atrophies, due to the criteria of Claus and Aschoff in addition local atrophies of midbrain structures, pons, width of the cervical medulla.

Group 2:
— T2 weighted signal relation between Globus pallidum and other nuclei of the basal ganglia, mainly N. caudatus
    1. intensity of Pallidum < than other nuclei (normal)
    2. intensity of Pallidum > / = other nuclei (pathological)
    3. periventricular hyperintense rim on T2 weighted images
— Atrophy of N. caudatus, characterized by a flattening of the caput and an increased width of the frontal horns.

Group 3:
— T1 weighted hyperintensities related to basal ganglia, subcortical regions and N. dentatus cerebelli
— T2 weighted hypointensities in these locations.

Group 4:
— T2 weighted hyperintense lesions related to the basal ganglia
— intensity of Pallidum related to other nuclei (evaluation due to criteria of group 2).

Results

Group 1:

Table 1. Pars compacta width (Fig. 1 and 2)

Normal controls	n = 15	2,75 (0,37)
MP	n = 13*	1,92 (0,34)
PS	n = 9	2,59 (0,33)
SP	n = 6	2,40 (0,42)

(Standard deviations in brackets)

* In 4 patients width of pars compacta could not be measured because of signal restoration within the posterolateral part of the zona reticulata (Rutledge et al., 1987)

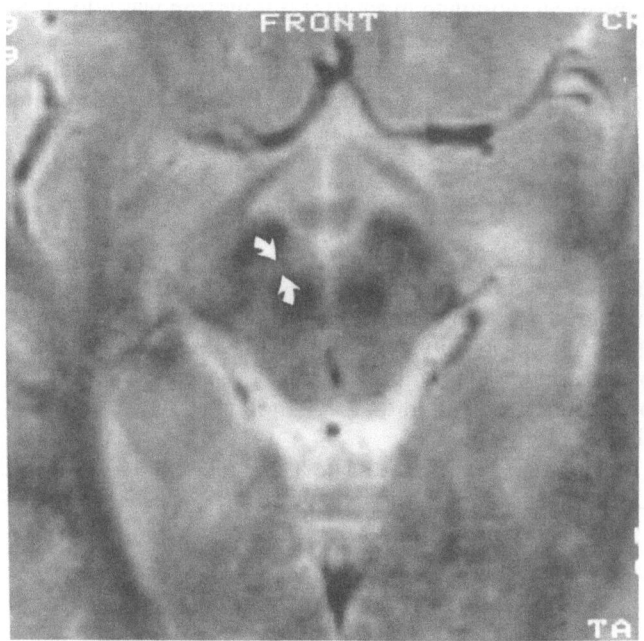

Fig. 1. 42 years old healthy woman. T2-weighted axial image (TR / TE 2500/30). Normal midbrain structures, regular Pars compacta width (white arrows)

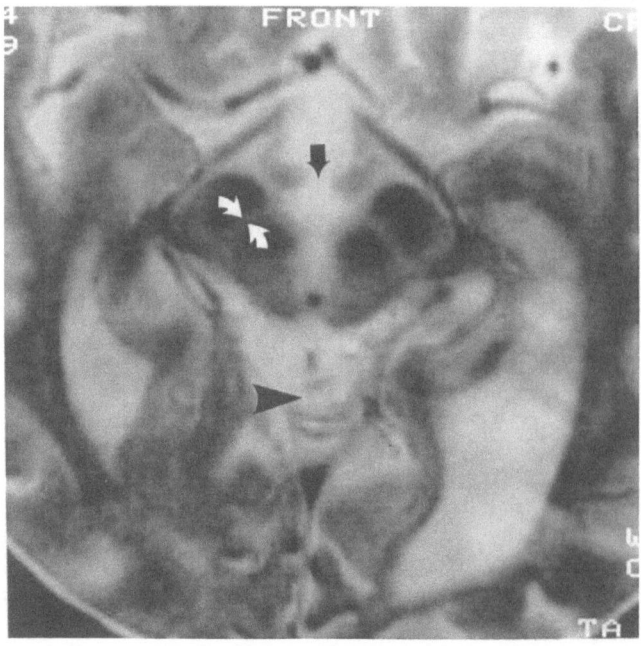

Fig. 2. 39 years old woman with OPCA, clinical symptoms since childhood. T2 weighted axial image (TR / TE 2500/30). Pars compacta width reduced (white arrows), atrophies of the upper vermis (black arrow head) and the cerebral peduncles with wide interpeduncular cistern, long distance between the Corpora mamillaria (black arrow)

Table 2. Signal relation Pallidum / Putamen

		Normal	Pathol.
MP	n = 13	7	4
PS	n = 9	4	5
SP	n = 6	5	1

Table 3. Atrophies

		Vermis	Hemisph.	Pons	Midbrain	Medulla
MP	n = 13	0	5	0	0	0
PS	n = 9	4	6	4	3	2
SP	n = 6	0	2	0	0	0

Table 4. T2 weighted hyperintense lesions

		Basal gang.	White m.
MP	n = 13	4	4
PS	n = 9	2	1
SP	n = 6	3	4

Group 2:

Table 5. (n = 5)

	Normal	Pathological
Signal relation Pallidum / Other nuclei	3	2
Shape of caput N. caudati	0	5
Periventricular zone	4	1

T2 weighted hyperintense spots occured in 1 patient.

Group 3:

Table 6. (n = 5)

	T1 weighted hyperintensities	T2 weighted hypointensities
Basal ganglia	5	5
Subcort. reg.	1	0
N. dentatus cereb.	3	1

2 patients had T2 weighted hyperintense lesions periventricular and / or in the semiovale center.

Group 4:

Table 7. (n = 38)

	found	not found
T2 weighted hyperintensity in basal ganglia	16	22
Pathological signal relation Pallidum / other nuclei	0	38

One patient who underwent a follow up examination after 13 months of penicillamine therapy showed a remarkable improvement of basal ganglia lesions.

3 patients had patchy T2 weighted white matter lesions.

Discussion

Group 1

Compared with a normal group, Pars compacta width was reduced in patients with M. Parkinson, Parkinson diseases and secondary parkinsonism. This pattern is supposed to be caused by selective loss of neuromelanin containing cells and increased ferritin deposition, while normal Pars compacta is free from ferritin (Braffman et al., 1988). Significant reduction of Pars compacta width we found only in MP patients, who presented a 31% reduction of Pars compacta width. Pars compacta width reduction of PS patients and SP patients was 6% and 13%. This may be related to the relatively high mean age and the longer duration of illness of the MP subgroup, compared with the others (Duguid et al., 1986).

Stern et al. (1989) observed a significant reduction of Pars compacta width in Parkinson patients as well as in patients with PD (Duguid et al., 1986). Pathological signal relations between Pallidum and Putamen were observed more frequently in patients with PD than in patients with MP (Drayer, 1989). Also in our few patients belonging to the subgroups MP and PD, the last one showed more often this pathological feature. Infratentorial atrophies concerning pons and medulla oblongata were only observed in certain PD, like OPCA and spinocerebellar atrophy. These atrophy patterns have been found out as specific morphological signs which may even precede clinical disabilities (Duguid et al., 1986).

T2 weighted white spots and areas of ischemical infarction occured more often in patients with vascular risk factors and SP. Further they correlated with age of the patients.

Group 2

Flattening of the caput N. caudati, as it appeared in all 5 of our patients is a common finding with Chorea Huntington, which has already been described by CT scan (Starkstein et al., 1989). Pathological ferritin deposition, as we saw in 2 of our patients, can only be revealed by high field MRI. As ferritin abnormalities may occur before N. caudatus atrophies, they can support a clinically unclear diagnosis as we observed in one of our patients who suffered from a Westphal type of Chorea Huntington, imitating Parkinsonism (Rutledge et al., 1987). Periventricular hyperintense rim on T2 weighted images, seen in one of our patients seems to be an unspecific sign, which has not only been found in choreatic patients, but also in patients suffering from M. Alzheimer (Duguid et al., 1986; Fazekas et al., 1987).

Group 3

Fahr disease may result from genetically determined errors in calcium or parathormone metabolism, or more frequently, from parathyreoprivic operations. In the latter group latency between operation and onset of the first clinical symptoms may be up to 30 years (König, 1985). MRI visualizes the different states of calcification, which appear T1 weighted hyperintense when consisting mainly of an albuminoid matrix and T2 weighted hypointense when containing calcium or other mineral ions without binding proteins (Lang et al., 1989). Thus the stage of the disease may be determined.

Group 4

T2 weighted hyperintense basal ganglia lesions correspond to edema, spongy changes and cavitation. Probability of morphological proof of these changes rises with duration of disease. Therapy related reversibility was found in one case. Pathologic cerebral presence of copper does not cause any T1 or T2 shortening. Thus no T2 weighted abnormal hypointensities could be identified in the basal ganglia.

Conclusion

Concerning basal ganglia diseases MRI improves information about:
1. Differential diagnosis between MP and PD or other diseases, involving the extrapyramidal system, which is of importance for the planning of therapeutic strategies (Drayer, 1989).
2. Stage of disease and thus prognosis of impending disabilities, as clinical symptoms may occur with latency to morphological changes. Further different levels of tissue alteration can be identified. It will be the purpose of follow-up investigations to find out whether one of the stages of Fahr disease features a reversible state when calcium and phosphate metabolism is recompensated. This assumption is supported by our experiences with M. Wilson patients. M. Wilson parallels Fahr disease as well concerning etiological mechanisms as clinical manifestations (König, 1985). In M. Wilson response to penicillamine therapy could be monitored by MRI (Prayer et al., 1990).

3. Biochemical processes in extrapyramidal diseases are displayed by the different features of mineral deposition. Thus a better understanding of mechanisms of basal ganglia diseases in vivo and as a consequence of new therapeutic regimes is initiated.

References

Braffman BH, Grossmann RI, Goldberg HI, Stern MB, Hurtig DB, Bilaniuk LT, Zimmermann RA (1988) MR imaging of Parkinson disease with spin-echo and gradient-echo sequences. AJNR 9: 1093–1099

Claus D, Aschoff J (1982) Evaluation of infratentorial atrophy by computed tomography. J Neurol Neurosurg Psychiatry 45: 979–983

Drayer BP (1989) Magnetic resonance imaging and extrapyramidal movement disorders. Eur Neurol 29: 9–12

Duguid JR, De La Paz R, De Grott J (1986) Magnetic resonance imaging of the midbrain in Parkinson's disease. Ann Neurol 20: 744–747

Fazekas F, Chawluk JB, Alavi A, Hurtig HI, Zimmermann RA (1987) MR signal abnormalities at 1.5 T in Alzheimer's dementia and normal aging. AJNR 8: 421–426

Flügel KA (1985) Parkinson-Syndrome im Rahmen multisystemischer Erkrankungen des Zentralnervensystems. In: Schnabert G, Auff E (Hrsg) Das Parkinson-Syndrom. Klinik, Neuropathophysiologie, Therapie, klinische Schwerpunkte. Roche, Wien, S47–51

Huckmann MS (1982) Computed tomography in the diagnosis of degenerative brain disease. Radiol Clin North Am 20: 169–183

König P (1985) Zur Psychopathologie des Fahr'schen Syndroms. Habilitationsschrift. Bibliomed, Melsungen

Lang C, Huk W, Pichl J (1989) Comparison of extensive brain calcification in postoperative hypoparathyroidism on CT and NMR scan. Neuroradiology 31: 29–32

Prayer L, Wimberger D, Kramer J, Grimm G, Imhof H (1990) Cranial MRI in Wilson's disease. Neuroradiology 32

Rutledge JN, Hilal SK, Silver AJ, Defendini R, Fahn S (1987) Study of movement disorders and brain iron by MR. AJNR 8: 397–411

Starkstein SE, Folstein SE, Brandt J, Pearlson GD, Mc Donnell A, Folstein M (1989) Brain atrophy in Huntington's disease. A CT-scan study. Neuroradiology 31: 156–159

Stern MB, Braffman BH, Skolnick BE, Hurtig HI, Grossmann RI (1989) Magnetic resonance imaging in Parkinson's disease and parkinsonian syndromes. Neurology 39: 1524–1526

Authors' address: D. Wimberger, MD, Neurological Clinic, University of Vienna Medical School, Lazarettgasse 14, A-1090 Vienna, Austria

J Neural Transm (1991) [Suppl] 33: 141–147
© by Springer-Verlag 1991

Extrapyramidal disturbances after cyanide poisoning (first MRT-investigation of the brain)

B. Messing

Psychiatrisches Landeskrankenhaus, Wiesloch, Federal Republic of Germany

Summary. A 29 year old student of chemistry took 50 ml of a 1% potassium cyanide solution (500 mg) in attempted suicide. He became comatose, mydriatic and was admitted to hospital in an apneic state. He woke up after seven hours and developed Parkinsonism in the following weeks. This regressed slowly in the second month after the poisoning apart from dysarthria, bradykinesia of the upper limbs and very brisk monosynaptic reflexes. Three weeks after the intoxication, CCT was largely normal, and there was CSF-dense hypodensity in both putamina after five months. Sharply delimited signal elevation in T2 corresponding to the two putamina was detected in the MRI eight weeks and five months after ingestion of the poison.

Cyanides occur in abundance in nature but is also found in technology and even forms in humans during the metabolism of some wellknown medical preparations (for example the antihypertensive agent Ni-Pruss, the not longer used cytostatic agent Laetrile and the antidot against thallium poisoning Berliner Blue).

The high cyanide concentration of some insects protect them from birds because of the repelling taste. Also the eucalyptus leaves content a high amount of cyanide salts, which the coala bear uses to maintain its energy balance as it has a natural tolerance to cyanide. Some humans also have a tolerance to cyanide salts, as the desintegration of cyanide salts in the human stomach is ph-dependent and that, in case of anacidity of the gastric juice, fewer cyanide salts are cleaved so that less cyanogen can be absorbed. This is probably why Grigorii Jefimowitsch Rasputin survived various potassium cyanide poisonings.

Already in ancient times cyanides were used to kill evil-doers of the Isis by an infusion of apricot kernels. Recently Bogdahn et al./Würzburg described a presumed cyanide poisoning with elder.

In technics cyanides are still used for disinfection of ships and buildings.

The human brain (particularly the basal ganglia) is highly sensitive to cyanides. Cyanide salts are rapidly absorbed, and cyanide radicals inactivate cytochrome oxidase, the terminal enzyme in respiratory electron transport

Table 1. Cyanide poisoning. Pathological findings in human brain

Schmorl (1920)	Globus pallidus
Edelmann (1921)	(as in salvasan poisoning or in Wilson's disease)
Pentschew (1958)	Putamina
Ule and Pribilla (1962)	Putamina
Braico et al.(1979)	Putamina
Kim et al. (1982)	Globus pallidus
Uitti et al. (1985)	Putamina

chain which utilizes the oxygen derived from the dissociation of oxyhemo-globin. This finally leads to effective suspension of all cellular respiration. There is no essential difference between the neurological effects of NaCN, KCN, HCN and CNC. The mortality of cyanide salts is 95% by a low lethal dose of 1 mg/kg of body weight and the LD 50 for orally ingested potassium cyanide is 3 mg/kg of body weight. As death mostly occurs in less than 3 minutes from respiratory paralysis, cyanide salts are often used in suicide attempts such as the mass suicide of 900 fatalists in Jonestown. As a result of the fast and reliable effectiveness, for a long time only cerebro-pathological findings were known. The first publications by Schmorl (1920) and by Edel-mann (1921) (Table 1) described the same case and showed the pathological changes in the globus pallidus similar as the case described by Kim (1982) does. Schmorl mentioned that cyanogen damages the same nuclei of the basal ganglia as salvarsan (Ehrlich HATA 606) and Wilson's disease. In the other cases, which are listed in Table 1 the putamen was involved.

Experiments on animals produce similar, but largely extended findings in the brain. Table 2 shows, that the putamen and the globus pallidus are damaged in all of the animals with exception of the putamen of the cat. In our case we observed and treated a 29-year old chemistry student who swallowed 50 ml of a 1% potassium cyanide solution corresponding to 500 mg potassium cyanide in an attempt of suicide. The Fig. 1 shows the symptomesand signs in relation to the time, which appeared after the poisoning. He quickly became comatose and his pupils were mydriatic. In this condition he was found by his parents. Gastric lavage was carried out in the emergency ambulance and since he was apnoeic, intubation and artificial respiration were administered. This state of unconsciousness disappeared after 7 hours. Creatinase and leukocytes increased during the first days, reflux oesophagitis and vomiting developed and later noticeable oculogyrias.

After 3 weeks the tongue swelled, somewhat later the patient developed dysarthria, his speech slowed down and was accompanied by voice chances. After four and a half weeks, dysphagia occured accompanied by tongue motility disturbance. The patient could not longer put food between his teeth when eating. Hypomimia and hypokinesia followed this stage which were completed by further extrapyramidal motor disturbances. A gait disturbance developed, characterized by the fact, that the patient could only move with deeply bent knees and a stiff back as well as pronounced rigidity of the cervical muscles and the upper limbs connected with a bradydiadochokinesia of

Table 2. Pathological findings in animal experiments

Author	Animal	Damaged regions
Meyer (1933)	Monkey	Globus pallidus, Putamina, N. caudatus
Ferraro (1933)		N. niger, Cortex, Cerebellum
Hurst (1940)		
Hicks (1944)	Rat	Globus pallidus, Putamina, N. caudatus
		N. niger, Cortex cerebri, Cerebellum
Haymaker et al. (1952)	Dog	Globus pallidus, Putamina, N. caudatus, N. niger, Cortex cerebri
Brierley (1975)	Rhesus	Globus pallidus, Putamina, N. caudatus, Cortex cerebri
Funata (1984)	Cat	Globus pallidus, N. niger, Corpus callo sum

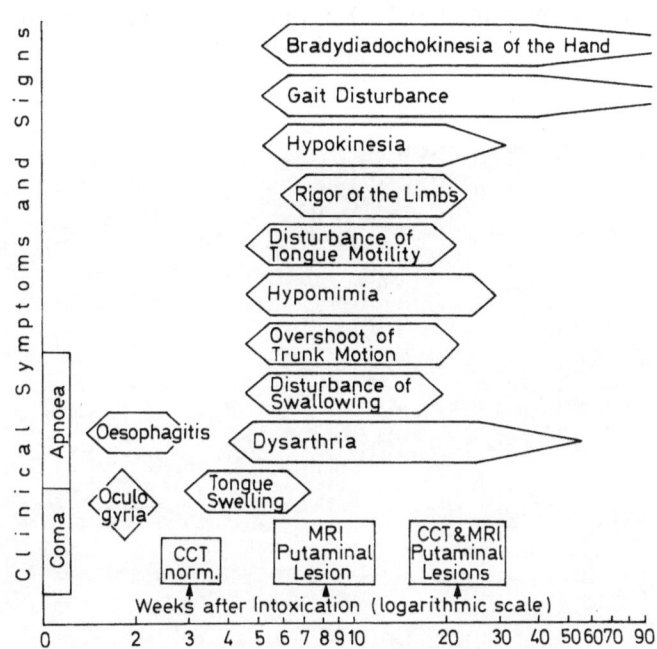

Fig. 1. Timetable of emerging and disappearing of symptomes and signs

the hands which prevented the patient from playing the organ. During the next 20 weeks most of these extrapyramidal disturbances improved under treatment of 6 mg Biperidin/day and later Carbidopa. After 25 weeks there was only a dysarthria as well as the gait disturbance described and the bradydiadochokinesia in the hands. 2 years later, the voice was normal and the speech disorder has completely improved. The bradydiadochokinesia has nearly disappeared and only disturbed the patient in writing or fine drawings. In this times he could play the organ again. ENT findings were always normal and no neurophysiological disorders occured. The individual reflexes were increased and remained so after intoxication, no pathological external reflexes

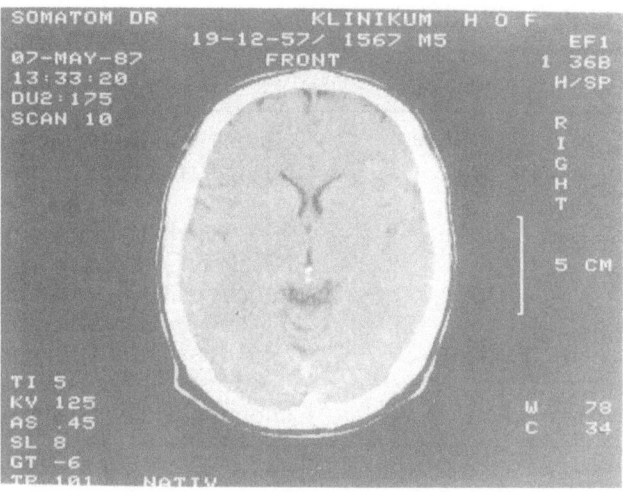

Fig. 2. Computerized tomography (CCT) 3 weeks after intoxication. Still no abnormalities in the basal ganglia

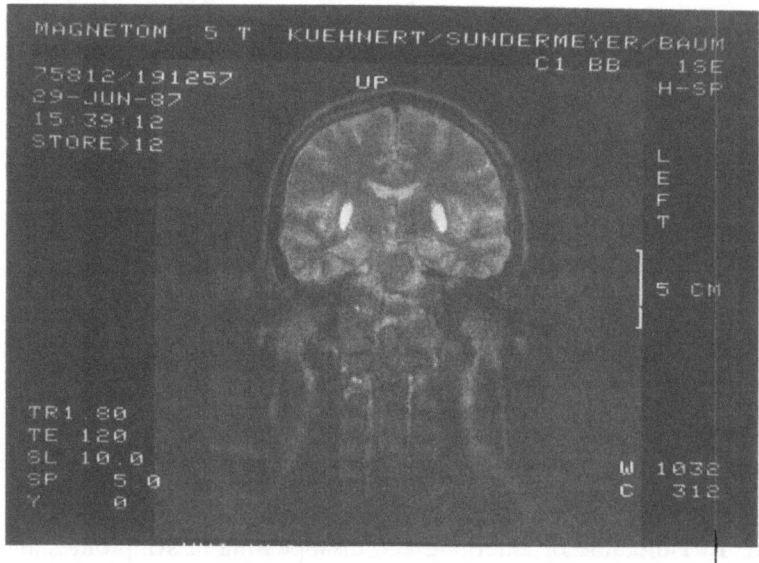

Fig. 3. Magnetic resonance imaging (MRI). Coronal sections 2 month after ingestion of poison. In T2 weighting, there is symmetrically arranged signal elevation which corresponds to the putamina

occured at any time. The patient is now successfully pursuing his studies again.

The CCT examination (Fig. 2), carried out 3 weeks after intoxication showed normal findings of the basal ganglia and small ventricles. Due to the occurence of extrapyramidal disturbances, a MRT was carried out after 8 weeks (Fig. 3) and revealed a signal elevation in the T2 weighted images in both putamina. In the MRT after 22 weeks this finding had not changed, but in the CCT (Fig. 4) CSF-isodense putamina could be represented and a clear

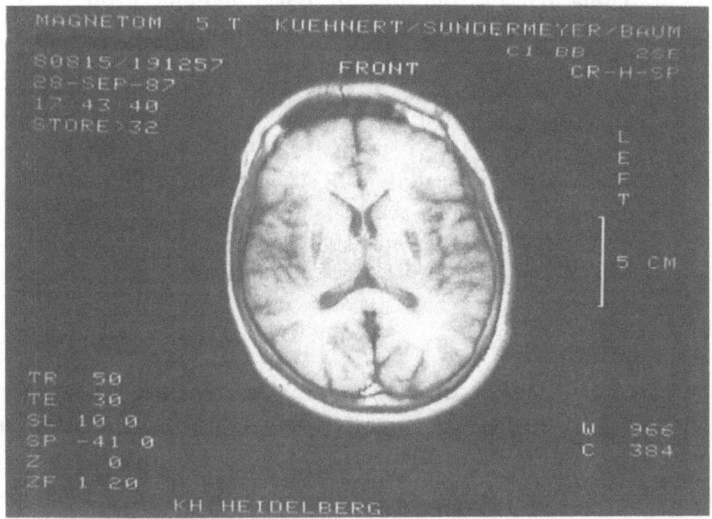

Fig. 4. CCT 5 month after intoxication. CSF-dense putamina. Relatively wide ventricles and pronounced subarachnoid spaces

Table 3. Cyanide poisoning in CCT and MRI

Braico et al. (1979)	Laetrile poisoning 1-year old child survived 71 h	CCT: bilat. putaminal infarction unilat. cerebellar hematoma Autopsia: putaminal infarction
Finelli (1981)	Ca-cyanide attempted suicide 30 year old man	CCT: bilat. symmetrical infarction of the globus pallidus unilat. cerebellar hematoma
Jacobs (1984)	HCN vapor Fireman	CCT: normal on 2. and 4. day after poisoning
Messing and Storch (1988)	KCN attempted suicide 29-year old man	CCT: bilat. CSF-isodense putamina MRT: high signal intensity in T2 of both putamina

expansion of the lateral ventricles as well as a dilatation of the subarachnoid space.

In literature (Table 3) only a lateral ventricle dilatation was found in the CCT findings of a one-year old child after Laetrile intoxication. The child survived the poisoning for almost 3 days. However, the post mortem revealed infarction in both putamina and in one cerebellar hemisphere.

The case of Finelli in 1981 is the first in which a bilateral lesion in the globus pallidus could be detected in the CCT.

Jacob's case — apparently a slight intoxication at a fireman with hydrocyanide acid and clinical normalization — showed normal findings on the 2nd and 4th day after intoxication.

Finally in our case a lesion covering the entire putamen was detected for the first time in CCT and also in MRT.

Table 4. CCT-findings in basal ganglia after carbonmonoxide and methanol poisoning

Carbonmonoxide	
Sawada et al. (1980)	
Destee et al. (1985)	Globus pallidus
Jaeckle and Nasrallah (1985)	Globus pallidus
Methanol	
Aquilonius et al. (1978, 1980)	Putamen
Mc Lean et al. (1980)	Putamen

The symptomes and signes described in our case are probably not specific for cyanide but an expression of the disturbed function of the nuclei caused by cyanide. Schmorl noticed as early as 1920 (Table 1) that the cerebral localisation after cyanide poisoning is identical with that after arsphenamine poisoning and Morbus Wilson.

A survey of the damaged sites after carbon monoxide and methanol poisoning (Table 4) points to the affinity of these substances for the globus pallidus and the putamen. While carbon monoxide apparently favors the globus pallidus, methanol has an affinity to the putamen.

On the other hand, there seems to be species-related differences as cats were the only animals examined which did not have lesions of the putamen while this nucleus in monkeys, dogs and rats were damaged (Table 2). However, based on the existing literature it is not possible to provide more information on this aspect. This information can perhaps be obtained in the future with the use of CCT and MRT studies in survivors. The same applies to the question whether the different poisons damaging the basal ganglia produce a special symptome constellation. At any rate, the new imaging procedures on the brain offer an excellent opportunity to develop a correlation between Parkinson's disease and the particularly affected nuclei in the different kinds of poisonings discussed above.

References

Aquilonius SM, Asmark H, Enoksson P, Lundberg PO, Moström U (1978) Computerized tomography in serve methanol intoxication. Br Med J 2: 929–930

Aquilonius SM, Bergström K, Enoksson P, Hedstrand U, Lundberg PO, Moström U, Olsson Y (1980) Cerebral computer tomography in methanol intoxication. J Comput Assist Tomogr 4: 425–428

Braico KT, Humbert JR, Terplan K, Lehotay JM (1979) Laetrile intoxication. Report of a fatal case. N Engl J Med 300: 238–240

Brierley JJB (1975) A comparison between the effects of profound hypotension, hypoxia and cyanide to the brain of M. Mulatta in primate models of neurological disorders. In: Meldrum BS (ed) Raven Press, New York, pp 116–128

Destee A, Courteville V, Devos PH, Besson P, Warot P (1985) Computing tomography and acute carbon monoxide poisoning. J Neurol Neurosurg Psychiatry 48: 281–282

Edelmann F (1921) Ein Beitrag zur Vergiftung mit gasförmiger Blausäure, insbesondere zu den dabei auftretenden Gehirnveränderungen. Dtsch Z Nervenheilkd 72: 259–287

Ferraro (1933) cit. from Hurst 1940

Finelli PF (1981) Case report changes in the basal ganglia following cyanide poisoning. J Comput Assist Tomogr 5: 775–756

Funata N, Song S-Y, Okeda R, Funata M, Higashino F (1984) A study of experimental cyanide encephalopathy in the acute phase-physiological and neuropathological correlation. Acta Neuropathol 107: 64–99

Haymaker W, Ginzler AM, Ferguson RL (1952) Residual neuropathological effects at cyanide poisoning: a study of the central nervous system of 23 dogs exposed to cyanide compounds. Milit Surg 111: 231–246

Hicks (1940) cit from Hurst 1940

Hurst EW (1940) Experimental demyelination of the central nervous system. I. The encephalopathy by potassium cyanide. Aust J Exp Biol Med Sci 18: 201–223

Jacobs K (1984) Erfahrungsbericht über die Anwendung von 4DMAP bei schwerer Blausäurevergiftung. Konsequenzen für die Praxis. Zentralbl Arbeitsmed 34: 274–277

Jaeckle RS, Nasrallah HA (1985) Major depression and carbon monoxide-induced parkinsonism: diagnosis, computerized axial tomography and response to L-Dopa. J Nerv Ment Dis 173: 503–508

Kim YH, Foo M, Terry RD (1982) Cyanide encephalopathy following therapy with sodium nitroprusside. Arch Pathol Lab Med 106: 392–393

McLean DR, Jacobs H, Mielke BW (1980) Methanol poisoning. A clinical and pathological study. Ann Neurol 8: 161–167

Messing B, Storch B (1988) Computer tomography and magnetic resonance imaging in cyanide poisoning. Eur Arch Psychiatr Neurol Sci 237: 139–143

Meyer A (1933) Experimentelle Vergiftungsstudien. III. Über Gehirnveränderungen bei experimenteller Blausäurevergiftung. Z Ges Neurol Psychiatr 143: 333–348

Pentschew A (1958) Blausäurevergiftung. In: Lubarsch VO, Henke F, Rössle R (Hrsg) Handbuch der speziellen Pathologischen Anatomie und Histologie, Bd Nervensystem, 2. Teil, Bandteil B. Springer, Berlin Göttingen Heidelberg, 13. 2149–2156

Sawada Y, Takahashi M, Ohashi N (1980) Computerized tomography as an indication of long term outcome after acute carbon monoxide poisoning. Lancet i: 783–784

Schmorl G (1920) Gehirn bei Blausäurevergiftung. Münch Med Wochenschr 67: 913

Uitti RJ, Rajput AH, Ashenhurst EM, Rozdilsky B (1985) Cyanide-induced parkinsonism: a clinico-pathologic report. Neurology 35: 921–925

Ule G, Pribilla O (1962) Hirnveränderungen nach Cyankalivergiftung mit protrahiertem (intervallären) klinischen Verlauf. Acta Neuropathol (Berlin) 1: 406–410

Authors' address: Dr. B. Messing, Psychiatrisches Landeskrankenhaus, Heidelberger Straße 1, D-W-6908 Wiesloch, Federal Republic of Germany

J Neural Transm (1991) [Suppl] 33: 149–155
© by Springer-Verlag 1991

Stroke: evaluation of long-term rehabilitation effects

A. V. Ungern-Sternberg[1], **M. Küthmann**[1], and **G. Weimann**[2]

[1]Department of Internal Medicine II, and [2]Department of Internal Medicine I,
Weserbergland-Klinik, Höxter, Federal Republic of Germany

Summary. A planned prospective documentation of the course of rehabilitation of 303 stroke patients was undertaken using the Bathel-Index as a measure of basic everyday functions and the Guttman-Scale as a measure of complex activities of daily living. These were determined at the beginning of rehabilitation, after an average of 7 weeks of in-patient treatment and one year following the stroke. Four patterns in the course of rehabilitation could be differentiated. The causes of the differing functional results were investigated. Besides a positive spontanious progress of the underlaying disease with an early reparation of the neurological deficits it is the premorbid status, the overprotection of the physically disabled and the determinative cognitive and mental functions that decide the long term fate of stroke patients.

Introduction

Primary goal of rehabilitation after stroke is a complete recovery of neurological deficits to facilitate social and occupational reintegration. The evaluation of rehabilitation success should be based onto objective criteria (Carey and Posavac, 1982; Feigenson, 1982; Gresham, 1986; Seal and Davis, 1987). Dombovy (1986) reviewed the literature and in most instances described an inappropriate design and a lack of standardized measurement in stroke outcome studies. A few indices measuring basic motoric skills and activities of daily living (ADL) have found general acceptance, especially the Barthel — index (Anderson et al., 1979; Blumendhal and Koch, 1981; Carey and Posavac, 1982; Chino et al., 1988; Forer and Miller, 1980; Granger et al., 1979, 1989, 1975; Kotila et al., 1984; Lindmark, 1988; Wade et al., 1983). This ADL index however does not adequately describe cognitive and other complex abilities as pointed out by Carter et al. (1988) and Freed and Wainapel (1983) and Feigenson et al. (1977a, b) who has described the linkage between cognitive and perceptive dysfunction and its influence on general outcome. The present study investigated the long-term functional state of CVA victims and describes their pattern on recovery.

Method

The course of recovery of 303 stroke patients was documented in a prospectively planned multicenter study analysing the medical history, neuropsychological status and rehabilitation treatments of these patients. To evaluate outcome the Barthel-index (basic independency) and Guttman's activity scale (more advanced activities of daily living) were assessed at predetermined intervals. The subsequent analysis is based on index measurements at admission, discharge and one year after rehabiltation. To evaluate the influence of age of 24 to 59 years (X = 50.1 years; n = 174) and 60 to 84 years (X = 69.7 years; n = 129).

Statistics

The statistical analysis was calculated with the SPSSPC 3.0 software package. A $p < 0.05$ was considered to be significant. Nominal data were analyzed using the CHI^2 — test.

Results

Independent of age four patterns of recovery — measured by Barthel-index and Guttman scale — can be described accounting for 98% of all patients (Table 1a, b). Generally, older patients have lower ADL levels (10 to 15%) but the pattern of recovery is similar for all ages. Two groups can be identified starting from comparable levels of dependency with significant improvements during rehabilitation but absolut dissimilar long-term functional development.

Group 1 shows a continous improvement of independency within the 12 month follow-up period while group 2 declines back to dependency levels seen at the begin of the rehabilitation treatment independent of the therapy during this period.

A third group (Group 3) accounting for 31% of the patients over 60 years and for 15% of the younger ones is basically independent at onset of rehabiliation with regard to the Barthel-index and only slightly impaired as measured by the Guttman scale and showing only clinically insignificant functional change during the observation period (Fig. 1a and b).

Group 4 consists of patients (15%) admitted with a high Barthel-index (average: 80 Pts) and with a minor loss of function during rehabilitation.

Summing the two indices (Fig. 2) the specific group pattern emerges with a leveling of functional capabilities for group 1, 2 and 4 during intensive rehabilitation. Analysis of the Barthel-index shows that the items "transfer", "walking" and "walking stairs" contribute the most to the achieved gains (Table 2).

Discussion

Based on Granger's (Delisa et al., 1982; Hayes and Carrol, 1986; Kinsella and Ford, 1980; Kotila et al., 1984; Lindmark, 1988; Tovack et al., 1984) description of severe dependence (i.e. Barthel-index below 60 points) (Granger

Table 1a. Barthel-score

Total	Admission		Discharge		Follow-up	
n = 303	Y	O	Y	O	Y	O
Group 1	75	70	90	84	93	88
Group 2	77	64	91	79	84	63
Group 3	100	100	100	100	99	99
Group 4	88	78	88	78	88	74

Y = 24–59 years; O = 60–84 years

Table 1b. Guttman-scale (%)

Total	Admission		Discharge		Follow-up	
n = 303	Y	O	Y	O	Y	O
Group 1	62	58	70	63	79	67
Group 2	63	54	70	63	68	58
Group 3	85	88	89	92	90	88
Group 4	74	69	74	63	75	63

Y = 24–59 years; O = 60–84 years

Table 2. Development of self-care

	Group 1			Group 2		
	Admission	Discharge	Follow-up	Admission	Discharge	Follow-up
Barthel-Score	70	84	88	64	79	63
Eating	58%	+8	+6	53%	+9	+9
Transfer	75%	**+18**	+3	66%	**+24**	**−26**
Personal hygiene	89%	+10	0	79%	+11	−14
Toilet	85%	+10	+1	72%	+12	−10
Bathing	7%	+15	+11	0%	+10	−7
Walking	64%	**+28**	+1	62%	**+23**	**−26**
Climb stairs	41%	**+26**	+19	38%	**+21**	**−11**
Dress	70%	+9	+5	57%	+7	−14
Rectal continence	96%	+4	0	98%	0	−5
Urinary continence	90%	+7	+1	91%	+4	−9

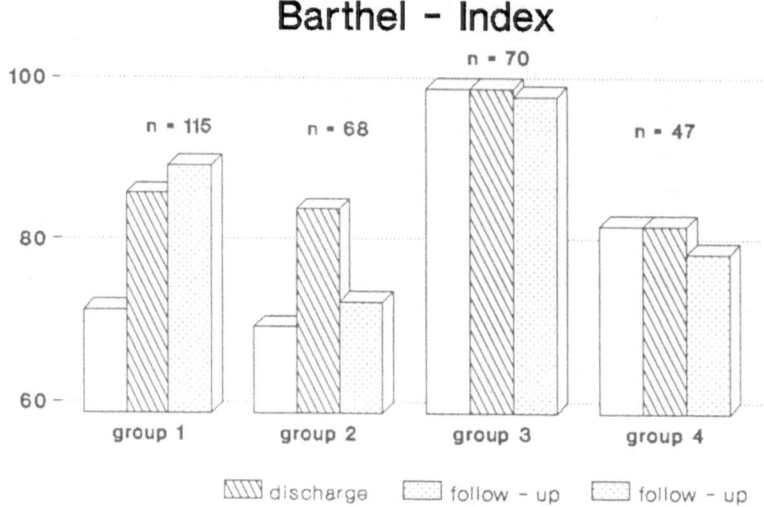

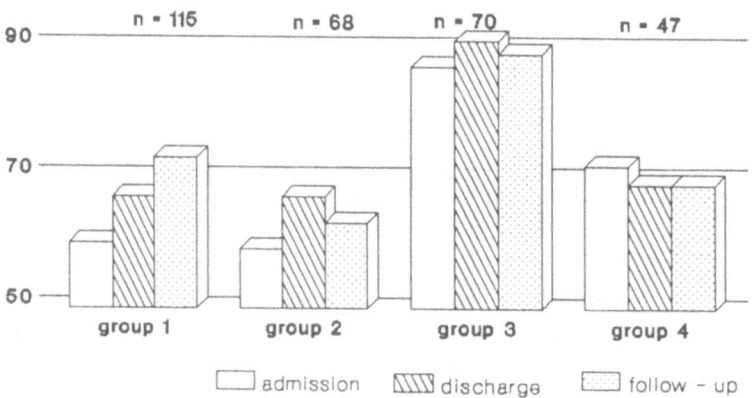

Fig. 1. Course of rehabilitation

et al., 1979) the observed patients had a rather high level of independency upon admission in our hospital regardless of their age and a similar admission intervall since the CVA. Feigenson observed a different recovery rate as a function of delay between CVA and rehabilitation (Feigenson et al., 1977b), a pattern that we did not see in our sample. The often cited early recovery of neurological deficit within the first three months after the incident is the most likely reason for independency at admission in the case of 30% of the younger and 15% of the older patients. To allocate resources accordingly to patients needs these patients would probably benefit more from an ambulatory rehabilitation. An explanation for the functional loss in Group 2 within the first year is the "overprotection" at home, especially for the older patients. This problem had been described by other authors (Garraway et al., 1980; Heine-

Barthel + Guttman - Index (%)

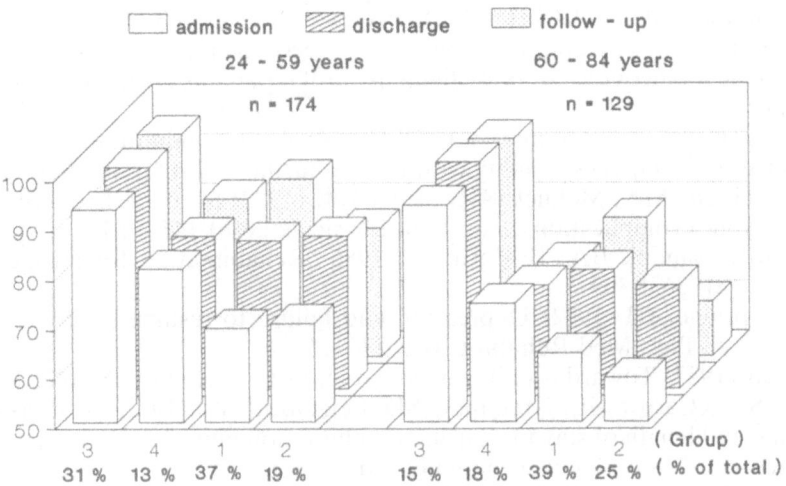

Fig. 2. Quality of live — Score

mann et al., 1987) as serious problem in stroke rehabilitation as well and deserves more attention in the therapeutic concepts to avoid a waste of medical and human efforts (Osberg et al., 1988; Reding and McDowell, 1989; Smith et al., 1981).

Typical risk factors of CVA, initial incontinence or aphasia as well as unconsciousness had no prevailing influence on outcome. High premorbid social activity levels as well as the general health correlated significantly with the chance of continous long-term improvement. These results add another dimension to earlier publications putting emphasis on the treatment of depression and cognitive dysfunction (Carter et al., 1988; Feigenson et al., 1977a; Freed and Weinapel, 1983; Feibel and Springer, 1982; Robinson et al., 1985). The described different pattern of recovery in CVA patients suggests that in about 25% of the cases a sufficient rehabilitation setting can be achieved through a well planned ambulatory rehabiliation care system.

Acknowledgements

In cooperation with the study group for Applied Social Gerontology University of Kassel. Supported by a grant of the Volkswagenwerk Foundation.

References

Anderson TP, Baldridge M, Ettinger MG (1979) Quality of care for completed stroke without rehabilitation: evaluation by assessing patient outcomes. Arch Phys Med Rehabil 60: 103–107

154   A. V. Ungern-Sternberg et al.

Blumenthal W, Koch M (1981) Leistungsbewertung und Wiedereingliederung Behinderter. Rehabilitation (Stuttg) 20(1): 8–12

Carey RG, Posavac EJ (1982) Rehabilitation program evaluation using a revised level of rehabilitation scale (LORS-II). Arch Phys Med Rehabil 63: 367–370

Carter LT, Oliveira DO, Duponte J, Lynch SV (1988) The relationship of cognitive skills performance to activities of daily living in stroke patients. Am J Occup Ther 42(7): 449–455

Chino N, Anderson TP, Granger CV (1988) Stroke rehabilitation outcome studies: comparison of a Japanese facility with 17 U.S. facilities. Int Disabil Stud 10: 150–153

Delisa JA, Mikulic MA, Melnick RR, Miller RM (1982) Stroke rehabilitation. Part II. Recovery and complications. Am Fam Physician 26(6) 143–151

Dombovy ML, Sandok BA, Basford JR (1986) Rehabilitation for stroke: a review. Stroke 17(3): 363–369

Feibel JH, Springer CJ (1982) Depression and failure to resume social activities after stroke. Arch Phys Med Rehabil 63(6): 276–277

Feigenson JS (1982) Toward a uniform assessment of "outcome". Stroke 13: 873–876

Feigenson JS, McCarthy ML, Greenberg SD, Feigenson WD (1977a) Factors influencing outcome and length of stay in a stroke rehabilitation unit. Part 2. Comparison of 318 screened and 248 unscreened patients. Stroke 8(6): 657–662

Feigenson JS, McDowell FH, Meese P, McCarthy ML, Greenberg SD (1977b) Factors influencing outcome and length of stay in a stroke rehabilitation unit. Part 1. Analysis of 248 unscreened patients — medical and functional prognostic indicators. Stroke 8(6): 351–356

Freed MM, Wainapel SF (1983) Predictors of stroke outcome. Am Fam Physician 28(5): 119–123

Forer SK, Miller LS (1980) Rehabilitation outcome: comparative analysis of different patient types. Arch Phys Med Rehabil 61: 359–365

Garraway WM, Akhtar AJ, Hockey L, Prescott RJ (1980) Management of acute stroke in the elderly: follow-up of a controlled tria. Br Med J 281(6244): 827–829

Granger CV, Dewis LS, Peters NC, Sherwood CC, Barrett JE (1979) Stroke rehabilitation: analysis of repeated Barthel index measures. Arch Phys Med Rehabil 60(1): 14–17

Granger CV, Greer DS, Liset E, Coulombe J, O'Brien E (1975) Measurement of outcomes of care for stroke patients. Stroke 6(1): 34–41

Granger CV, Hamilton BB, Gresham GE, Kramer AA (1989) The stroke rehabilitation outcome study. Part II. Relative merits of the total Barthel index score and a four-item subscore in predicting patient. Arch Phys Med Rehabil 70(2): 100–103

Gresham GE (1986) Stroke outcome research. Stroke 17(3): 358–360

Gresham GE, Phillips TF, Labi ML (1980) ADL status in stroke: relative merits of three standard indexes. Arch Phys Med Rehabil 61(8): 355–358

Hayes SH, Carroll SR (1986) Early intervention care in the acute stroke patient. Arch Phys Med Rehabil 67: 319–321

Heinemann AW, Roth EJ, Cichowski K, Betts HB (1987) Multivariate analysis of improvement and outcome following stroke rehabilitation. Arch Neurol 44(11): 1167–117

Kinsella G, Ford B (1980) Acute recovery from patterns in stroke patients: neuropsychological factors. Med J Aust 2: 663–66

Kotila M, Waltimo O, Niemi ML, Laaksonen R, Lempinen M (1984) The profile of recovery from stroke and factors influencing outcome. Stroke 15(6): 1039–1104

Lindmark B (1988) Evaluation of functional capacity after stroke with special emphasis on motor function and activities of daily living. Scand J Rehabil Med [Suppl] 21: 1–40

Novack TA, Satterfield WT, Lyons K, Kolski G, Hackmeyer L, Connor M (1984) Stroke onset and rehabilitation: time lag as a factor in treatment outcome. Arch Phys Med Rehabil 65(6): 316–319

Osberg JS, McGinnis GE, DeJong G, Seward ML, Germaine J (1988) Long-term utilization and charges among post-rehabilitation stroke patients. Am J Phys Med Rehabil 67(2): 66–72

Reding MJ, Mc Dowell FH (1989) Focused stroke rehabilitation programs improve outcome. Arch Neurol 46: 700–701

Robinson RG, Starr LB, Lipsey JR, Rao K, Price TR (1985) A two-year longitudinal study of poststroke mood disorders. In-hospital prognostic factors associated with six-month outcome. J Nerv Ment Dis 173: 221–226

Seale C, Davies P (1987) Outcome measurement in stroke rehabilitation research. Int Disabil Stud 9(4): 155–160

Smith DS, Goldenberg E, Ashburn A, Kinsella G, Sheikh K, Brennan PJ, Meade TW, Zutshi DW, Perry JD, Reeback JS (1981) Remedial therapy after stroke: a randomised controlled trial. Br Med J (Clin Res) 282(6263): 517–520

Wade DT, Skilbeck CE, Hewer RL (1983) Predicting Barthel ADL score at 6 months after an acute stroke. Arch Phys Med Rehabil 64: 24–28

Authors' address: Prof. Dr. A. von Ungern-Sternberg, Abteilung für Innere Medizin II, Weserberglandklinik, D-W-3470 Höxter, Federal Republic of Germany

J Neural Transm (1991) [Suppl] 33: 157–161
© by Springer-Verlag 1991

Medical educational and functional determinants of employment after stroke

H. Bergmann[1], **M. Küthmann**[1], **A. v. Ungern-Sternberg**[1], and **V. G. Weimann**[2]

[1]Department of Internal Medicine II and [2]Department of Internal Medicine I, Weserbergland-Klinik, Höxter, Federal Republic of Germany

Summary. To evaluate the medical educational and functional determinants of employment after stroke a total of 204 Patients were assessed for functional changes during rehabilitation and after one year. Additionally the educational and social background of each patient was documented. One year after discharge from the rehabilitation unit 11.3% of the patients worked full-time and 2.7% part time. Another 14.7% had the decision pension versus employment still pending and 0.5% were classified as unemployed. 70.8% received regular retirement plan payments or disability pensions. Generally a similar level of functional capabilities can be observed after one year for working and non-working patients except for manual dexterity. Early admission to rehabilitation (within the first 12 weeks) favours return to work. A high school degree qualifying for university entrance and a well paid and better qualified profession as well are correlated with a higher percentage of patients regaining their employment. An other determining factor of employment after CVA is the physical requirement of the former work. Most of the working patients (92%) had been either transfered onto a job suited for an handicapped people or their work plan had been restructed accordingly.

Introduction

Most outcome studies investigating the rehabilitation of stroke patients have centered on activity of daily living (ADL) parameters (Anderson et al., 1979; Chino et al., 1988; Feigenson et al., 1979; Feigenson, 1982; Forer and Miller, 1980; Gresham et al., 1979; Hayes and Carroll, 1986; Kinsella and Ford, 1980; Lewis, 1986; Robinson et al., 1985; Sarno et al., 1985; Wade et al., 1983) or therapy and prevention (Asplund et al., 1981; Hemmingsen et al., 1982). Only a few publications (Becker et al., 1986; Howard et al., 1985; Lawrence and Christie, 1979; Singer, 1987; Tan, 1983; Terent, 1983) have documented the employment status after the cerebrovascular accident (CVA). Different health and social security systems make comparison of such data from various countries even more difficult to evaluate. Employment after CVA has been reported in 50% in one publication (18) while other authors described employment rates between 10% and 22% (Becker et al., 1986; Terent, 1983; Singer, 1987).

This study evaluated the medical, functional and educational background of younger stroke patients employed one year after rehabilitation for a cerebrovascular accident.

Method

204 patients (Mean age: 50.7 years, s.d. 7.6) were assessed for functional changes during rehabilitation and after one year (n = 174). Additionally the educational and social background of each patient was documented. To be included in the study patients had to be younger than 59 years at the time of the CVA assuming a minimum regular retirement age of 60 years.

Data analysis was mainly based on descriptive statistics.

Statistical significance of nominal data was calculated using the CHI^2-test. A value of $p < 0.05$ was considered significant.

Results

One year after discharge from the rehabilitation unit 11.3% of the patients worked full-time and 2.7% part-time. Another 14.7% had the decision pension versus employment still pending and 0.5% were classified as unemployed. All others received regular retirement plan payments or disabilty pensions.

The functional status of working versus retired/disabled patients is summarized in Table 1. Generally a similar level of functional capabilities can be observed after one year for working and non-working patients except for manual dexterity. The analysis of the intervall between CVA and admission into the rehabilitation unit showed a clear difference between these groups with early admission more frequently for the patients that eventually work again:

— Admission within in 8 weeks : 40% versus 11.3%
— Admission within 12 weeks : 46.7 versus 35.2%
— Admission after 6 months   : 0.0% versus 33.4%

Table 1. Important functional status parameters and employment

Percent	Admission		Discharge		Follow-up	
	W	R/D	W	R/D	W	R/D
Walking stairs	61.5	42.5	100	68.8	76.2	80.8
Manual dexterity	26.7	1.1	57.1	12.0	63.6	26.8
Comprehension	100.0	79.8	100.0	84.8	95.5	81.8
	n = 204		n = 204		n = 174	

W working; *R/D* retired or disability pension

Professional status before CVA

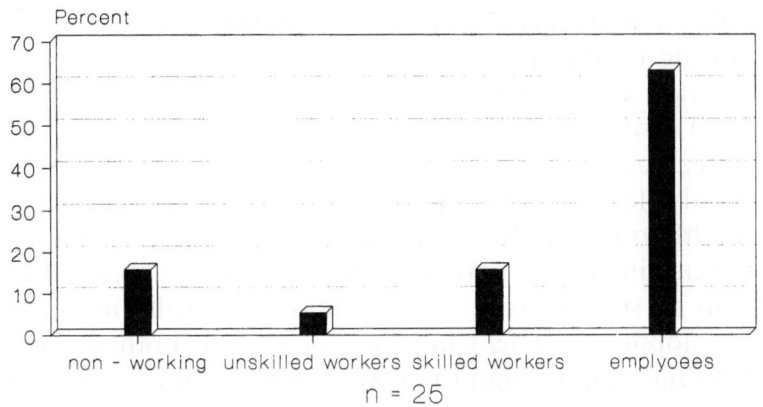

n = 25

Fig. 1. Employment after one year

Premorbid educational and professional status

A total of 19.4% of the patients had a high school degree qualifying for university entrance. Grouped by working versus non-working patients after one year the working patients dominated with 18.2% versus 1,2% (p < 0.05). Analysis of the professional status prior to stroke shows a higher percentages of well paid and better qualified stroke patients regaining their employment (p < 0.05) (Fig. 1).

Another determing factor of employment after CVA is the physical requirement of the former work (p < 0.05) with 40.9% of the employed cases having a job that can be performed sitting while the retired patients had only in 10.1% such working conditions. Factors such as the type of industry, size of company or duration of employment did not influence the chance of returning to work.

About 92% of the working patients had been either transfered onto a job suited for handicapped people or their work place had been restructered accordingly.

Discussion

The interaction of rehabilitation effects and return to full employment after CVA are generally difficult to assess as many internal (i.e. motivation, job satisfaction) and external factors (i.e. financial security, qualification, unemployment rate) influence the final outcome. While some of these conditions cannot be changed by rehabilitation measurements others can be influenced to a certain degree. Comparison of the functional status of patients showed that the manual dexterity of the affected side differs tremendously between working and retired patients (63.6% versus 26.8%) while other factors like comprehension and walking have improved in almost all patients.

Generally, retired patients still make large gains after discharge from the rehabilitation unit.

One possible way to raise the rather low return to work rate might be a closer monitoring of the improving functional capabilities after discharge from rehabilitation at set intervals (i.e. every 3 months within the first year after the CVA) as there is possible a misconception about the intrinsic improvement and associated work chance. Often the judgement of employment chances is based solely on the medical discharge records of the rehabilitation unit which might no reflect the true capabilities at a later point in time. Secondly, the monitoring of the patient's status during rehabilitation and afterwards needs to be standardized to get reliable and valid information about functional developments. This study supports the hypothesis that early admission to rehabilitation (within the first 12 weeks) obviously favours return to work. However the degree of "confounding" factors such as influence of high motivation on admission chances cannot be evaluated by our data and remains subject of further investigation.

References

Anderson TP, Baldridge M, Ettinger MG (1979) Quality of care for completed stroke without rehabilitation: evaluation by assessing patient outcomes. Arch Phys Med Rehabil 60: 103–107

Asplund K, Liliequist B, Fodstad H, Wester PO (1981) Long-term outcome in cerebrovascular disease in relation to findings at aortocervical angiography. A 12-year follow up. Stroke 12: 307–313

Becker C, Howard G, McLeroy KR, Yatsu FM, Toole JF, Coull B, Feibel J, Walker MD (1986) Community hospital-based stroke programs: North Carolina, Oregon, and New York. II. Description of study population. Stroke 17: 285–293

Chino N, Anderson TP, Granger CV (1988) Stroke rehabilitation outcome studies: comparison of a Japanese facility with 17 U.S. facilities. Int Disabil Stud 10: 150–153

Feigenson JS, Gitlow HS, Greenberg SD (1979) The disability oriented rehabilitation unit — a major factor influencing stroke outcome. Stroke 10: 5–8

Feigenson JS (1982) Toward a uniform assessment of "outcome". Stroke 13: 873–876

Forer SK, Miller LS (1980) Rehabilitation outcome: comparative analysis of different patient types. Arch Phys Med Rehabil 61: 359–365

Gresham GE, Phillips TF, Wolf PA, McNamara PM, Kannel WB, Dawber TR (1979) Epidemiologic profile of long-term stroke disability: the Framinghamstudy. Arch Phys Med Rehabil 60: 487–491

Hayes SH, Carroll SR (1986) Early intervention care in the acute stroke patient. Arch Phys Med Rehabil 67: 319–321

Hemmingsen R, Mejsholm B, Boysen G, Engell HC (1982) Intellectual function in patients with transient ischaemic attacks (TIA) or minor stroke. Long-term improvement after carotid endarterectomy. Acta Neurol Scand 66: 145–159

Howard G, Till JS, Toole JF, Matthews C, Truscott BL (1985) Factors influencing return to work following cerebral infarction. JAMA 253: 226–232

Kinsella G, Ford B (1980) Acute recovery from patterns in stroke patients: neuropsychological factors. Med J Aust 2: 663–666

Lawrence L, Christie D (1979) Quality of life after stroke: a three-year follow-up. Age Ageing 8: 167–172

Lewis NA (1986) Functional gains in CVA patients: a nursing approach. Rehabit Nurs 11: 25–27

Robinson RG, Starr LB, Lipsey JR, Rao K, Price TR (1985) A two-year longitudinal study of poststroke mood disorders. In-hospital prognostic factors associated with six-month outcome. J Nerv Ment Dis 173: 221–226

Sarno MT, Buonaguro A, Levita E (1985) Gender and recovery from aphasia after stroke. J Nerv Ment Dis 173: 605–609

Singer F (1987) Zur Rehabilitation der cerebralen Apoplexie-Ergebnisse anhand von 156 Patienten. Rehabilitation 26 (1): 1–7

Tan ES (1983) Stroke rehabilitation — Singapore experience. Ann Acad Med Singapore 12 (3): 373–376

Terent A (1983) Medico — social consquences and of stroke in a Swedish Community. Scand J Rehabil Med 15 (4): 165–171

Wade DT, Skilbeck CE, Hewer RL (1983) Predicting Barthel ADL score at 6 months after an acute stroke. Arch Phys Med Rehabil 64: 24–28

Authors' address: Dr. H. Bergmann, Department of Internal Medicine II, Weserberglandklinik, D-W-3470 Höxter, Federal Republic of Germany

Subject Index

Supplementum 32

P. Riederer and M. B. H. Youdim (eds.)

Amine Oxidases and Their Impact on Neurobiology

Proceedings of the
4ᵗʰ International Amine Oxidases Workshop,
Würzburg, Federal Republic of Germany, July 7–10, 1990

This book describes the most recent research in the field of monoamine oxidase (MAO), diamine oxidase and semicarbazide-sensitive amine oxidase. MAO and its subtypes have gained enourmous interest in the treatment of a variety of disorders, like depression syndrome Parkinson's disease and possibly dementia of Alzheimer type. Recently selective and reversible MAO-A inhibitors (brofaromine, moclobemide) have been developed. They are safe and easy to handle in clinical practice and show reduced side effects compared to the early irreversible and unspecific MAO-I's. In addition the irreversible and selective MAO-B inhibitor has been shown to prolong life expectancy of patients with Parkinson's disease. Therefore, this book contains additionally to about 36 articles on MAO and its inhibitors, a number of chapters dealing with oxidative stress resulting from radical processes derived in part from deamination. In contrast to other books on the subject, the role of catechol-O-methyltransferase, sulfation and up-take-processes are discussed in detail. This is of particular interest as such an overall view allows a closer insight on neuronal and extraneuronal metabolizing processes.

For the reader the publication is of great interest in so far as the book has been published within four months after hand in of manuscripts. Therefore, the book is a timely written publication of todays research in the field.

1990. 110 figures. XII, 491 pages.
Soft cover DM 240,–, öS 1680,–
Reduced price for subscribers to
"Journal of Neural Transmission":
Soft cover DM 216,–, öS 1512,–
ISBN 3-211-82239-9

Prices are subject to change without notice

Springer-Verlag
Wien New York

Supplementum 31

H. Glossmann (ed.)

New Therapeutic Uses of Calcium Channel Blockers

This publication presents an overview on new therapeutic uses of calcium channel blockers and future trends for calcium antagonists. The areas covered are tissue-selectivity, myocardial protection, renal effects of calcium antagonists and actions on the peripheral and central nervous system, including cognitive function as well as neuronal repair. Authors, which are competent researchers in the respective fields present own novel data and cover the most recent literature.

Springer-Verlag
Wien New York

1990. 21 figures. VII, 74 pages.
Soft cover DM 65,–, öS 450,–
Reduced price for subscribers to
"Journal of Neural Transmission":
Soft cover DM 58,50, öS 405,–
ISBN 3-211-82200-3

Prices are subject to change without notice